Vortex Dynamics in the High-Temperature Superconductors $YBa_2Cu_3O_7$ and $Bi_2Sr_2CaCu_2O_{8+\delta}$ in Low- and High-Dissipative Transport Measurements

by

Patrick Voss-de Haan

uniuer
sität⊕
mainz

Mainz, Germany, 1999

D77

Vortex-Dynamik in den Hochtemperatur-Supraleitern $YBa_2Cu_3O_7$ und $Bi_2Sr_2CaCu_2O_{8+\delta}$ in niedrig- und hochdissipativen Transportmessungen

—

Vom Sprung im Glas

—

Dissertation

zur Erlangung des Grades

„Doktor der Naturwissenschaften"

(Dr. rer. nat.)

am Fachbereich Physik der

Johannes Gutenberg-Universität

in Mainz

von

Patrick Voss-de Haan

geb. in Hamburg

univer sität⊛ mainz

Mainz, 1999

D77

Referent: Prof. Dr. H. Adrian

Korreferent: Prof. Dr. G. Schönhense

Dekan: Prof. Dr. Th. Walcher

Tag der Einreichung: 11. 10. 1999

Tag der mündlichen Prüfung: 22. 12. 1999

Patrick Voss-de Haan

VORTEX DYNAMICS IN THE HIGH-TEMPERATURE SUPERCONDUCTORS $YBa_2Cu_3O_7$ AND $Bi_2Sr_2CaCu_2O_{8+\delta}$ IN LOW- AND HIGH-DISSIPATIVE TRANSPORT MEASUREMENTS

ibidem-Verlag
Stuttgart

Die Deutsche Bibliothek - CIP-Einheitsaufnahme:

Ein Titeldatensatz für diese Publikation ist bei
Der Deutschen Bibliothek erhältlich

∞

Gedruckt auf alterungsbeständigem, säurefreien Papier
Printed on acid-free paper

ISBN: 3-89821-024-3

© *ibidem*-Verlag
Stuttgart 2000
Alle Rechte vorbehalten

Printed in Germany

Contents

List of Figures

List of Tables

*Every sentence that I utter must be understood
not as an affirmation, but as a question.*

NIELS BOHR

The important thing is not to stop questioning.

ALBERT EINSTEIN

1 Introduction

The discovery of high-temperature superconductors (HTSC) is most commonly considered a particularly important achievement on the road to technical applications based on superconductivity. Yet, from the point of view of solid state physics an even more significant aspect of HTSC's may be the opportunity to investigate a variety of highly interesting physical mechanisms and phenomena related to superconductivity. These include the dimensionality and the dynamics of the vortex system, as well as topics of a more general nature, such as the existence of a glassy state of the vortex ensemble and its properties. The notable advantage of high-T_c superconductivity lies in the possibility to fine-tune the properties of the system by precisely controlling the experimental conditions of temperature, magnetic field, and current density as well as material parameters such as the chemical composition.

In this work the vortex dynamics of two of the most prominent superconducting compounds, $YBa_2Cu_3O_7$ (YBCO) and $Bi_2Sr_2CaCu_2O_{8+\delta}$ (BSCCO), were investigated by transport measurements at low and high dissipation levels. While both compounds are layered, cuprate superconductors with similar crystal structures and critical transition temperatures ~90 K, there exist crucial distinctions in their superconducting properties. In the context of the experiments to be discussed the most important difference is the changed interlayer coupling and, as a consequence, the altered dimensionality of the vortex systems in YBCO and BSCCO, which leads to distinct characteristics of the respective vortex dynamics in both the low- and the high-dissipative regime.

The first part of this work, comprised of Chaps. 2 through 4, presents the analyses of experiments at low dissipation levels. Low dissipation properties of superconductors are relevant to general physical aspects, such as the existence of a truly superconducting state with vanishing linear resistivity as opposed to flux creep mechanisms resulting in finite dissipation even at the lowest current densities. Furthermore, they are important with regard to technical applications, which usually demand minimal power dissipation. The relevant theoretical basis—general concepts of superconductivity and vortices in type II superconductors as well as dimensionality and different possible states of the flux-line system—is summarized in Chap. 2.

Chapter 3 concentrates on the influence of a changed coupling between the superconducting CuO_2 layers on the properties of $Bi_2Sr_2CaCu_2O_{8+\delta}$ in transport measurements of samples with systematically varied oxygen stoichiometry. Generally, altering the chemical composition of a superconductor has become an important tool in adapting the material to the specific needs of an experiment or application. The substitution of atoms in the crystal lattice and the change of the concentration of interstitial oxygen have been used to control such parameters as the transition temperature or the critical current density. In the present experiment specifically the aspect of the vortex system's dimensionality changing with the oxygen content of the samples was investigated in current-voltage characteristics. The systematic change of the behavior in this series of BSCCO films, which can be attributed to the altered flux-line tension, allows a more confident interpretation of the experiments than for a single sample.

Due to the lower electronic anisotropy of YBCO its vortex dynamics display particular differences from those found in BSCCO. Especially the more three-dimensional character of YBCO results in the formation of a low-temperature phase exhibiting characteristics, which have repeatedly been attributed to a vortex glass (VG) state. In Chap. 4 the results of a new experimental technique, utilizing extremely long measurement bridges, are presented. These striplines of lengths exceeding 100 mm offer the opportunity to extend the electric-field sensitivity of transport measurements by two to three orders of magnitude at the lower end which is crucial for the investigation of the vortex glass state. A glass scaling analysis of this extended electric-field range as well as an adapted and refined analysis of the crossover current of the vortex glass transition—in agreement with a number of further analyses—shed a new light on the properties of a vortex glass and question the relevance of a number of previously published results.

Transport measurements in the high-dissipative regime are at the focus of the second part of this work. While these processes had long been largely neglected, there has been an increasing effort in this area of superconductivity during the last years. Properties of superconductors at high dissipation levels are potentially interesting to technical applications dealing with high current densities, such as superconducting current-fault limiters, which have recently been developed and employed. Also superconducting high-field magnets and short-time current storage systems, which during normal operation show basically no dissipation but must not be destroyed in the case of a malfunction, are related to such high-dissipation vortex dynamics. Furthermore, the vortex dynamics in high dissipation processes, i.e. at high vortex velocities, are relevant to fundamental questions of superconductivity, for example the understanding of quasi-particle states and thermal processes in HTSC's. One of the most peculiar mechanisms is a vortex instability leading to a discontinuity in the resistivity of a type II superconductor, manifested as a voltage jump in current-voltage characteristics. Chapter 5 briefly reviews the theory of Larkin and Ovchinnikov and its extension by Bezuglyj and Shklovskij describing the mechanism of a vortex instability in the flux-flow region at high vortex velocities.

The first observation of this phenomenon in the BSCCO system is the subject of Chap. 6. Due to its low magnetic irreversibility field and the large flux-flow regime this system is an excellent candidate for the investigation of the Larkin-Ovchinnikov instability. Experimental deviations from the theoretical predictions previously observed in other systems (primarily low-temperature superconductors) were also found in BSCCO, yet the complete data can successfully be explained in terms of the extended Bezuglyj-Shklovskij model, which accounts for inevitable effects of quasi-particle heating in the process of the instability. Thus, it was possible to extract quasi-particle scattering rates for BSCCO in a wide range of temperatures below the super-conducting transition. Furthermore, the anisotropy of the vortex instability in this system was investigated and is well described by a quasi two-dimensional model for the vortex system.

Although there exist some vital differences to the situation in BSCCO, for YBCO a similar instability in current-voltage characteristics had recently been reported. In order to control and analyze the role of thermal effects in this phenomenon several films of YBCO were prepared on different substrate materials with varying heat-transfer capabilities. As discussed in Chap. 7, for the investigated samples the ob-tained quasi-particle scattering rates indicate a strong dependence on electron-electron interaction at higher temperatures close to the superconducting transition whereas at lower temperatures the contribution of electron-phonon scattering is found to dominate. The quantitative analysis of the experimental data revealed sur-prisingly good agreement with the Bezuglyj-Shklovskij theory despite the lack of an extended flux-flow regime preceding the voltage jump which distinguishes the YBCO from the BSCCO system. The chapter concludes with the observation of a phenomenon related to this distinction: a heretofore unreported direct correlation of the voltage instability with the vortex glass phase was found throughout the entire temperature and magnetic-field range in YBCO. The origin of this correlation is yet unclear. The possibility of a vortex glass ensemble depinning and the influence of vortex-vortex interactions in the glass phase are discussed.

With the quasi-DC measurements (i.e. current pulses of about 1 s) used for obtaining the current-voltage characteristics in Chaps. 6 and 7, it is impossible to gain information about the time evolution of the vortex instability. Thus, a technique for the investigation of the vortex instability in the microsecond range was developed and is presented in Chap. 8. A short-pulse, arbitrary-waveform current source spe-cifically designed to suit the requirements of such experiments allowed the recording of time-resolved current voltage characteristics on the same samples used in the pre-vious chapter. Besides offering an additional tool in the analysis of thermal effects, this technique revealed an evolution of the instability clearly extended over time. Interestingly, the interval of time needed for the rise of the measured voltage to the saturation value (close to the normal state resistivity) appears to be independent of temperature and magnetic field. However, above the vortex glass transition the rise time changes drastically and the actual instability vanishes, thus confirming the correlation of the vortex dynamics in the low-dissipative regime of the vortex glass and the high-dissipative regime of the vortex instability. Tentative explanations for this unpredicted phenomenon are offered based on the dynamic vortex mass and long entanglement relaxation times.

2 Vortex Dynamics at Low Dissipation Levels

This chapter presents a brief overview of the particular topics of superconductivity relevant to the analysis of the experimental results. After covering some basic aspects of superconductivity it concentrates on aspects relating to vortex dynamics in high-temperature superconductors. In particular the mechanisms of pinning, thermal activation, interlayer coupling, and vortex-vortex interactions will be considered and their effects on the dimensionality and the corresponding phases of the vortex system—such as a vortex glass—are discussed. For a more elaborate discussion of these and other more general topics of superconductivity the reader is referred, for example, to the introductory books by de Gennes and Tinkham [Genn66, Tink96]. A very detailed treatment of many special aspects of vortex dynamics in high-temperature superconductors, including an overview of experimental results, can be found in a number of reviews [Blatt94, Bran95, Cohe97].

2.1 Basic Concepts of Superconductivity

Two fundamental characteristics of superconductors are infinite conductivity, initially reported by Kamerlingh Onnes in 1911 [Kame11], and perfect diamagnetism, first observed in 1933 by Meissner and Ochsenfeld [Meis33]. However, both properties are destroyed at high temperatures and magnetic fields. The critical transition temperature T_c (in zero magnetic field) or alternatively the critical magnetic field $B_c(T)$ separate the normal and superconducting states in the phase diagram of a superconductor (cf. Sec. 2.6). The thermodynamic critical magnetic field can be calculated by means of the condensation energy of the superconducting state and the energy per unit volume necessary to expel the external magnetic field from the interior of the superconductor

$$f_n(T) - f_s(T) = \frac{B_c^2(T)}{2\mu_0}$$

(2.1)

Below B_c a sample will be in the Meissner state with basically no magnetic flux entering the superconductor while above B_c the magnetic field will penetrate the sample like any normal conducting material.

Phenomenologically, the relationship between electromagnetic fields and the current density in a superconductor has been described in a local approximation for the first time in 1935 by the London equations

$$\vec{E} = \frac{\partial}{\partial t}\left(\Lambda \vec{J}_s\right) \quad \vec{B} = -\text{curl}\left(\Lambda \vec{J}_s\right) \quad \Lambda = \mu_0 \lambda^2 = \frac{m_s}{n_s q_s^2} \tag{2.2}$$

where m_s, n_s, and q_s denote the mass, number density, and charge of the superconducting charge carriers, respectively [Lond35]. Combined with Maxwell's equation, (2.2) yields a solution $B \propto e^{-x/\lambda}$ corresponding to the existence of supercurrents within the characteristic penetration depth λ near the surface, which screen the interior of the superconductor from an external field.

In an attempt to improve this purely phenomenological approach the Ginzburg-Landau (GL) theory offered a conceptual explanation for superconductivity. Introduced in 1950, it concentrated on the superconducting electrons as defined in the London equations [Ginz50]. A complex pseudowavefunction ψ serves as an order parameter describing the local density of the superconducting electrons $n_s = |\psi(x)|^2$. The advantage of this approach is the ability to treat local variations in n_s. A derived differential equation for ψ based on two expansion coefficients α and β

$$\frac{1}{2m^*}\left(\frac{\hbar}{i}\nabla - e^* A\right)^2 \psi + \beta|\psi|^2 \psi = -\alpha(T)\psi \tag{2.3}$$

is basically analogous to a Schrödinger equation for a free particle with mass m^* and charge e^*. One of the important consequences of this model is the minimum distance ξ, called the GL coherence length, over which $\psi(x)$, and hence n_s, can vary without undue energy increase

$$\xi(T) = \frac{\hbar}{|2m^*\alpha(T)|^{1/2}} \tag{2.4}$$

Using the GL solution $\alpha(T) = -B_c^2\lambda^2 e^{*2}/m^*$ with $e^* = 2e$, $m^* = 2m$, and $\Phi_0 = h/2e$ one may rewrite this result

$$\xi(T) = \frac{\Phi_0}{2\sqrt{2}\pi B_c \lambda} \tag{2.4a}$$

As the penetration depth λ and the coherence length ξ both diverge as $(T_c - T)^{-1/2}$ the ratio $\kappa = \lambda/\xi$ of these two characteristic length scales is largely independent of temperature and defined by material properties. This Ginzburg-Landau parameter is related to the surface energy associated with domain walls between superconducting and normal material. For small values of $\kappa < 1/\sqrt{2}$ (type I superconductors) one obtains a positive surface energy, which is consistent with the magnetic field either completely penetrating or being completely expelled from the superconductor. However, for $\kappa > 1/\sqrt{2}$ (type II superconductors) the surface energy becomes negative and the response of a superconductor to an external magnetic field changes drastically as Abrikosov showed in 1957 [Abri57].

2.1.1 Type II Superconductors and Vortices

For sufficiently small external magnetic fields, type II just as type I superconductors will remain perfectly diamagnetic. However, due to the negative surface energy a type II system will favor an arrangement with many separate superconducting and normal domains maximizing the area of the phase boundary. Hence at a lower critical field $B_{c1} = B_c/\sqrt{2}\kappa$ magnetic flux will begin to continuously penetrate into the system resulting in an additional, vortex state particular to type II superconductors (cf. Fig. 2.1). Observed for the first time in 1937, in this Shubnikov phase superconducting and normal domains coexist and superconductivity can persist up to magnetic fields well above the thermodynamic critical field B_c [Schu37]. In fact, the sample will not become normal conducting until an upper critical field $B_{c2} = \sqrt{2}\kappa B_c$ is reached. As opposed to the intermediate state of a type I superconductor, where magnetic fields close to B_c can enter the system near the surface (due to sample geometry), magnetic flux will not enter a type II superconductor in the form of laminar domains. Instead, in order to minimize the free energy, the domains will subdivide until a quantum limit is reached, with each separate flux line carrying one magnetic flux quantum $\Phi_0 = h/2e = 2.07 \cdot 10^{-15}\,\mathrm{Tm}^2$. A vortex of supercurrents surrounds each flux line, which for an isolated vortex effectively shields the superconducting domain outside (with $B \approx 0$ as $r \to \infty$) from the normal core inside the flux line (where $B \approx 2B_{c1}$). The term vortex has therefore come to denominate the entire arrangement associated with a single flux line in a type II superconductor.

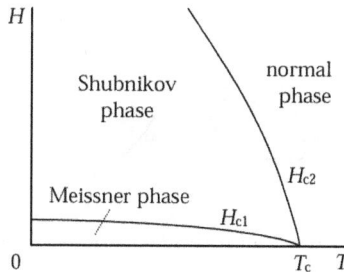

Figure 2.1: H–T phase diagram of a type II superconductor with the Meissner, Shubnikov, and normal phases separated by the critical fields $B_{c1}(T)$ and $B_{c2}(T)$.

7

In its lateral extension an isolated vortex is characterized by the penetration depth λ and the coherence length ξ. The former determines the behavior of the magnetic-field strength in the vicinity of the vortex and the latter is responsible for the suppression of the order parameter, illustrated in Fig. 2.2 for a typical type II superconductor with $\xi \ll \lambda$. Based on the GL theory one may approximate the value of the order parameter as a function of the distance r from the vortex center supposing circular symmetry [Tink96]

$$|\psi(r)| \approx \psi_\infty \cdot \tanh \frac{vr}{\xi} \qquad\qquad v \approx 1 \qquad\qquad (2.5)$$

One notes the sharp increase of $|\psi(r)|$ from zero at the center to almost ψ_∞ within the core radius $\sim\xi$. With the reasonable approximation of $|\psi(r)| \equiv \psi_\infty$ everywhere outside the core of the vortex, n_s becomes constant and the London equations may be used to treat the magnetic field and superconducting currents. Thus, for the dependence of the magnetic field on r one finds

$$B(r) = \frac{\Phi_0}{2\pi\lambda^2} K_0\left(\frac{r}{\lambda}\right) \qquad\qquad \xi \ll r \qquad\qquad (2.6)$$

where approximating the zero-th order Hankel function yields

$$B(r) \rightarrow \frac{\Phi_0}{2\pi\lambda^2} \cdot \sqrt{\frac{\pi}{2}\frac{\lambda}{r}} \cdot e^{-r/\lambda} \qquad\qquad r \rightarrow \infty \qquad\qquad (2.6a)$$

$$B(r) \approx \frac{\Phi_0}{2\pi\lambda^2} \cdot \left(\ln\frac{\lambda}{r} + 0.12\right) \qquad\qquad \xi \ll r \ll \lambda \qquad\qquad (2.6b)$$

From this description of the vortex structure it is possible to investigate the vortex-vortex interaction. Neglecting the core one derives the vortex free energy per unit length (line tension) for a single vortex due to the contributions of the field energy and the kinetic energy of the currents

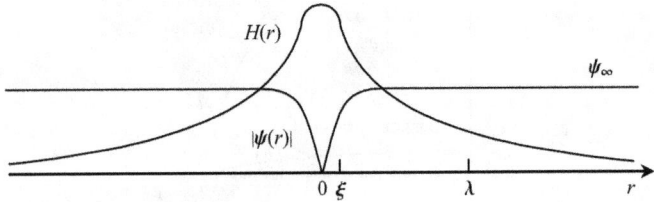

Figure 2.2:   Structure of an isolated vortex in a type II superconductor [Tink96].

$$\epsilon_1 = \frac{1}{2\mu_0} \int \left(\vec{B}^2 + \lambda^2 |\text{curl } \vec{B}|^2 \right) dS \qquad (2.7)$$

which, when combined with (2.6b) and cut off at $r = \xi$, results in

$$\epsilon_1 \approx \frac{1}{4\pi\mu_0} \left(\frac{\Phi_0}{\lambda} \right)^2 \ln \kappa = \frac{B_c^2}{2\mu_0} 4\pi\xi^2 \ln \kappa \qquad (2.7a)$$

As the free energy of the vortex is of the same magnitude as the condensation energy (2.1), except for the factor of $4 \ln \kappa$, this justifies the ansatz of ignoring the core for $\kappa \gg 1$. Furthermore, by substituting (2.6) into (2.7) one now obtains the interaction energy per unit length between two vortices separated by a distance r as

$$\epsilon_{12} = \frac{\Phi_0}{\mu_0} B(r) = \frac{\Phi_0^2}{2\pi\mu_0\lambda^2} K_0 \left(\frac{r}{\lambda} \right) \qquad (2.8)$$

which leads to analogous approximations

$$\epsilon_{12}(r) \rightarrow \frac{\Phi_0^2}{2\pi\mu_0\lambda^2} \cdot \sqrt{\frac{\pi}{2} \frac{\lambda}{r}} \cdot e^{-r/\lambda} \qquad r \rightarrow \infty \qquad (2.8a)$$

$$\epsilon_{12}(r) \approx \frac{\Phi_0^2}{2\pi\mu_0\lambda^2} \cdot \ln \frac{\lambda}{r} \qquad \xi \ll r \ll \lambda \qquad (2.8b)$$

For two vortices of the same orientation this results in a repulsive force; for vortices of opposite orientation, such as vortex-antivortex pairs (cf. [Miu98] and references therein) an attractive force arises. In non-negligible external magnetic fields all vortices can generally be considered to possess the same orientation and the mutual repulsion will result in a symmetric arrangement of the flux lines with a maximum intervortex spacing: the Abrikosov lattice [Abri57]. The system reaches a minimum energy for a triangular lattice and the distance between adjacent vortices is given approximately by $a_0 \sim \sqrt{\Phi_0 / B}$.

2.1.2 Vortex Solids, Flux-Line Melting, and Vortex Fluids

Like a crystal consisting of a lattice of atoms, the Abrikosov lattice—comprised of a regular arrangement of vortices with long-range order—can also be considered a solid. In a type II superconductor at sufficiently high temperatures, this vortex solid can also exhibit a first order melting transition from the Abrikosov lattice to a liquid state, which has been reported for single crystals of various HTSC's (e.g. YBCO and BSSCO, [Müll87, Gamm88]). This melting transition can be understood in terms of the increase of the thermal energy kT. The resulting vibrations of the individual vortices from their ideal positions in the lattice and the increase in vortex mobility

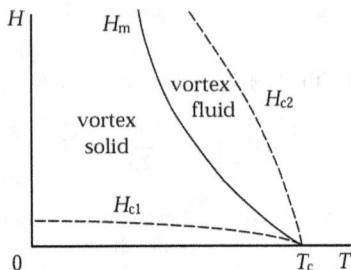

Figure 2.3: Schematic phase diagram of a type II superconductor with the melting transition. In the Shubnikov phase between H_{c1} and H_{c2} one distinguishes the solid state of the vortex system (e.g. an Abrikosov lattice), which exists at low magnetic fields and temperatures, and the liquid state without long-range order at higher H and T.

will finally destroy the regular lattice structure and lead to a liquid state at sufficiently high temperatures. While the flux lines are able to move individually, i.e. flow like the molecules in a liquid at temperatures $T_m < T < T_c$, the precise nature of the vortex dynamics in this flux-flow phase still depends on the actual value of the available thermal energy in relation to the vortex interaction energy. Near the melting temperature the vortex interaction in the liquid may be substantial and result in an increased viscosity while at higher temperatures it is negligible (cf. Secs. 2.2 and 2.4). The temperature dependence of the melting field is often also given as $B_m(T)$ or $H_m(T)$ in analogy to the temperature dependent critical fields $H_{c1}(T)$ and $H_{c2}(T)$. In order to define the melting transition in a theoretical treatment a Lindemann criterion [Lind10] is generally applied relating the displacement of flux lines from thermal fluctuations to a fraction of the lattice constant, $\langle u^2(T_m)\rangle_{th} \approx c_L^2 a_0^2$ [Blat94] with an experimental value of the Lindemann number $c_L \sim 0.2$–0.3 [Ma91, Seng91]. Hence, the magnetic-field dependence of the vortex lattice melting is understood as a consequence of the decrease of intervortex spacing at higher fields, where a smaller lateral perturbation will suffice to fulfill the Lindemann criterion.[1] A schematic vortex phase diagram is given in Fig. 2.3. Despite a noticeable material dependence of the melting line, its qualitative shape is very similar in most superconductors. In fact, it was shown that $H_m \propto (T\gamma)^{-2}$ [Tink96]. Therefore, melting of the flux-line lattice (FLL) is of much more interest in high-temperature rather than low-temperature superconductors as for the latter H_m and H_{c2} are usually undistinguishable due to their comparably small anisotropies γ and the low accessible temperatures.

---

[1] In principle, one expects an additional fluid region at extremely small magnetic fields below the solid regime and just above H_{c1}, where the intervortex distance becomes so large that each flux line can, in effect, be considered a single, free (non-interacting) vortex. Yet, this occurs at magnetic fields below the range of the experiments to be discussed and, moreover, in real samples this possibility is generally precluded by pinning.

2.1.3 High-Temperature Superconductors

The first high-temperature superconductor (HTSC), mixed phase Ba-La-Cu-O with a T_c = 35 K discovered by Bednorz and Müller in 1986 [Bedn86], was soon followed by compounds of similar structure but still higher transition temperatures: YBa$_2$Cu$_3$O$_7$ with $T_c \approx$ 93 K [Wu87], Bi$_2$Sr$_2$Ca$_{n-1}$Cu$_n$O$_{2n+4}$ with $T_c \leq$ 110 K [Maed88], Tl$_2$Ba$_2$Ca$_{n-1}$Cu$_n$O$_{2n+4}$ with $T_c \leq$ 130 K [Shen88], and finally HgBa$_2$Ca$_{n-1}$Cu$_n$O$_{2n+2}$ $T_c \leq$ 135 K [Schi93]. HTSC's are interesting objects for the study of superconductivity, especially since a number of physical mechanisms responsible for the intrinsic properties of these systems differ considerably from those of low-temperature superconductors (LTSC). For instance, in recent studies of the symmetry of the order parameter a $d_{x^2-y^2}$ symmetry was strongly favored [Kirt95, Harl95, Tsue97] but a complete absence of an isotropic s-wave pairing contribution has not been agreed upon. More importantly for this work, HTSC's possess high penetration depths and small coherence lengths resulting in very large Ginzburg-Landau parameters κ, which make them strong type II superconductors and excellent candidates for the study of vortex dynamics. Furthermore, it is possible to investigate superconductivity in these systems at comparatively high temperatures giving rise to a substantial influence of thermal activation of pinned vortices and a melting line clearly below the upper critical field with an extended liquid state. Likewise, the high-temperature range leads to sometimes quite strong superconducting fluctuation effects observable as a broadening of the resistive transitions (cf. Sect. 3.1.5). Finally, the intrinsic anisotropy due to the layered structure of HTSC's is also responsible for increasing the strength of the fluctuation effects and resulting in a material dependence as well as temperature and magnetic-field dependence of interlayer coupling and sample dimensionality. The following sections will discuss the theoretical consequences of these properties concentrating on the different possible phases of the vortex system in these superconductors and their dependence on temperature, magnetic field, dimensionality, and pinning.

2.2 Pinning and Activation Energy

The characteristics of the solid state, in which the interaction energy dominates, and of the liquid state, where thermal energy prevails, both also depend on the mechanism of pinning, which introduces disorder and strongly affects the dynamics of the vortex system. Supposing that, without an external current, the vortex ensemble is in the equilibrium state the net force due to the supercurrents of all other vortices must be zero. Hence, a superposition of a transport current of density J_{ext}

will simply result in a driving Lorentz force per unit length acting on any given vortex[2]

$$\vec{f}_{ext} = \vec{J}_{ext} \times \vec{\Phi}_0 \tag{2.9}$$

The vortices will begin to move perpendicular to the transport current and essentially induce an electric field of magnitude $\vec{E} = \vec{B} \times \vec{v}$. This resistive longitudinal voltage opposing J_{ext} will result in power dissipation for solid as well as liquid flux-line systems unless there exists a process preventing vortex motion, such as pinning of vortices by spatial inhomogeneities in the crystal lattice.

Possible pinning centers include impurities, voids, and deviations from stoichiometry on an atomic scale as well as grain boundaries, twin planes, dislocations, and artificial pinning sites such as columnar defects induced by heavy-ion irradiation. All these defects have in common that they represent deviations from the ideal crystal structure and destroy or at least reduce the superconducting properties in their vicinity. Due to the resulting suppression of the superconducting order parameter ψ at the sites of these defects, they present energetically favored locations for a vortex where less condensation energy is required for the reduction of the order parameter in the vortex core. Consequently, flux lines pinned to such defects will remain stationary as long as the pinning potential exceeds the available energy due to thermal activation and external transport currents. The activation energy needed to overcome the pinning barrier is related to the nature of the defect as well as to the properties of the vortex. Especially for thin films of high-temperature superconductors as opposed to single crystals, one generally finds a non-negligible density of pinning centers.

Pinning is often treated according to the model developed by Anderson and Kim, which suggests that thermal energy will allow bundles of vortices to jump between adjacent pinning sites [Ande62, Kim63, Ande64]. Without any external forces acting on the flux lines there will be no directional net movement of the entire ensemble due to the isotropic nature of this process. However, a nonzero current density will lead to a net jump rate in the direction of the Lorentz force. The term flux creep (FC) was coined for the gradual hopping movement of the vortices at low temperatures, with $E_{thermal} \ll E_{pinning}$, which explained the magnetization decay of type II superconductors observed on a logarithmic time scale. For higher temperatures, where $E_{thermal} \lesssim E_{pinning}$ and the dynamics are not dominated primarily by pinning, the vortex motion will resemble more closely a steady flow of vortices generally described as thermally activated flux flow (TAFF) [Tink88, Kes89] in analogy to the similar concept of free flux flow (FFF) [Tink64, Bard65].

The response of a sample to an external transport current allows to investigate the activation energy U via the resistivity $\rho(T)$ [Pals90]. Considering independent flux

---

[2] The influence of the Magnus force can be neglected in the context of this work.

lines or bundles of flux lines which move as a whole, the thermally induced jump rate for movement between adjacent pinning sites is given as

$$v = v_0 \cdot e^{-U/kT} \tag{2.10}$$

where v_0 is a characteristic attempt frequency. Due to the Lorentz force density $\vec{f} = \vec{J} \times \vec{B}$, a transport current will act on a flux bundle of volume V_c by moving it a distance r_p characterizing the range of the pinning potential. The work done by the current can be expressed as $U_J = JBV_c r_p$. In the direction of the Lorentz force this will lead to a reduction of the undisturbed pinning potential U_0 to an effective barrier potential $U = U_0 - U_J$; in the opposite direction the barrier potential will be increased to $U = U_0 + U_J$. Supposing a mean hopping distance L the resulting electric E field along the direction of the transport current is found

$$E = vB = Lv_{\text{eff}}B = 2Lv_0 B \cdot \exp\left(-\frac{U_0}{kT}\right) \cdot \sinh\left(\frac{JBV_c r_p}{kT}\right) \tag{2.11}$$

For small current densities one substitutes $\sinh x \approx x$ and recovers a linear dependence of E on J corresponding to ohmic dissipation. With a correlation length L_c of the flux-line bundle along its extension parallel to the c axis and in the 'amorphous limit', $V_c \approx a_0^2 L_c$ and $L \approx a_0$, one obtains the resistivity

$$\rho = \rho_0 \cdot \exp\left(-\frac{U_0}{kT}\right) \qquad \rho_0 = 2\Phi_0^2 \frac{v_0 L_c}{kT} \tag{2.12}$$

In order to be able to move, flux lines must surmount an intrinsic pinning barrier U_0, the activation energy. As its value is related to the average strength of a single pinning site but independent of the density of pinning centers, a study of the activation energy will be well-suited to analyze the character of pinning mechanisms in the comparative analysis of different samples and will be used extensively in Sec.3.1.5.[3] In this context, the influence of the dimensionality of a sample is of particular interest, as vortices in three-dimensional and two-dimensional systems should respond differently to pinning (cf. Sec. 2.3). A further consequence of (disordered) pinning in HTSC's, treated in Sec. 2.4, is the possibility of a true superconducting phase with vanishing linear resistivity, $\rho(J \to 0) = 0$, in a solid vortex phase at low temperatures, where infinite activation barriers effectively replace the finite barriers of the TAFF regime.

---

[3] Critical currents, which are also a common tool for investigating pinning mechanisms, relate to the overall pinning character in the sample rather than the individual pinning site as they also depend on the *density* of pinning centers, which may vary considerably for different samples.

2.3 Dimensionality in HTSC's

Regarding the anisotropic nature of HTSC's one may expect decisive differences in the dimensionality of vortex dynamics depending on the strength of the interlayer coupling. In the limit of strong decoupling, according to Kes's pancake model [Kes90, Clem91, Glaz91], a vortex can be considered an aligned stack of two-dimensional vortices: 'pancakes' which are only defined within the individual superconducting layers. These pancakes are weakly coupled by electromagnetic interaction and the Josephson effect [Jose62] and will be able to move more or less independently within each layer. The opposite extreme of strong coupling leads to the concept of a three-dimensional vortex (as in isotropic low-temperature type II superconductors) possessing a homogeneous structure along its axis and acting as a whole entity rather than separate pancake sections.

Obviously, the dimensionality of the vortices will thus influence the effective strength of overall pinning as a strongly coupled vortex may be pinned completely by a single defect while in the two-dimensional case only one of the pancake sections and its closest neighbors may be immobilized. More importantly, dimensionality can be crucial for the appearance of a variety of general physical phenomena closely related to a system's dimensionality, e.g. the existence of a vortex glass exhibiting true superconductivity only for a 3D system (see below) or a Kosterlitz-Thouless transition existing only in a 2D system [Kost73, Nels77, Halp79]. Temperature and magnetic field will modify the effective coupling strength of the flux lines leading to the coexistence of regions with two and three-dimensional behavior in the B-T phase diagram of some HTSC systems of intermediate anisotropy. HTSC's are interesting candidates for the investigation of dimensionality-dependent behavior especially because of the possibility of continuously adjusting the dimensionality between the limiting 3D and 2D cases by tuning their structural anisotropy.

2.3.1 Structural Anisotropy

Besides a generally high Ginzburg-Landau parameter κ resulting in extreme type II superconductivity, all HTSC's have in common the intrinsically layered crystal structure which gives rise to a strong anisotropy and a behavior distinctly different from the predominantly isotropic metallic LTSC's. The CuO_2 layers common to all HTSC's are believed to be the origin of the superconductivity in these systems. As a consequence, the charge carriers show quite different characteristics for movement within these planes as opposed to movement between adjacent planes. The aniso-tropic Ginzburg-Landau model, applying the ansatz of the isotropic theory to a set of n layers of two-dimensional systems, deals with this anisotropic behavior in terms of a reciprocal mass tensor $(1/m_{ij})$ which can be thought of as relating to the effective masses of the charge carriers, m_{ab} and m_c, in the principal crystal directions (inside the crystallographic ab plane and along the c axis) with the relatively small in-plane anisotropy between the *a* and *b* direction ignored. From the dependence of the coherence length on the effective mass m^* (2.4) it follows that ξ will assume different values ξ_{ab} and ξ_c in the *ab* plane and along the *c* axis, respectively [Tink96]

$$\xi_i \propto \frac{1}{\sqrt{m_i}} \qquad \lambda_i \propto \sqrt{m_i} \qquad i = a, b, c \qquad (2.13)$$

Thus, all superconducting properties and quantities relating to ξ or λ will be modified by a factor $\sqrt{m_c/m_{ab}}$. The anisotropy parameter γ of a superconducting system is generally defined by the ratio of the two effective masses and relates to coherence length and penetration depth as well as to lower and upper critical field in the *ab* plane and along the *c* axis

$$\gamma \equiv \sqrt{\frac{m_c}{m_{ab}}} = \frac{\xi_{ab}}{\xi_c} = \frac{\lambda_c}{\lambda_{ab}} = \frac{H_{c1\|c}}{H_{c1\|ab}} = \frac{H_{c2\|ab}}{H_{c2\|c}} \qquad (2.14)$$

With increasing anisotropy, $\gamma \geq 1$, the Josephson coupling between the segments of the same flux line in adjacent layers—and thus the three-dimensional nature of the originally isotropic system—will be reduced. In the limit of very large γ, with $m_c \gg m_{ab}$ and $\xi_c \ll \xi_{ab}$, one obtains the quasi-independent, two-dimensional systems of pancake vortices interacting within each layer rather than along the *c* axis. Best suited for the precise treatment of such extremely weakly coupled layered systems is the Lawrence-Doniach (LD) model, which explicitly considers the coupling of separate superconducting 2D layers via Josephson tunneling [Lawr70]. For inter-mediate values of γ the effective dimensionality of the flux-line system will depend on the ratio of the interlayer-coupling energy to the intraplane vortex-vortex interaction and to the thermal energy.

A controlled variation of γ may be achieved in two ways. First, one may change the crystallographic structure of the layered system, as has been done with super-lattices consisting of layers of normal and superconducting materials, where the thickness of the insulating layers separating the superconducting planes has been modulated. While this approach has yielded interesting results, e.g. an increase of anisotropy in YBCO-PrBCO superlattices [Bozo96], it generally suffers from a low reproducibility as it is very difficult to obtain these artificial crystal structures in sufficient quality. Alternatively, it is conceivable to alter the intrinsic properties of the superconducting material in order to influence the transport properties of *c* axis and *ab* plane separately, e.g. via a change of the charge-carrier density. This possibility will be a topic of Chap. 3.

2.3.2 Influence of Magnetic Field and Temperature

As the average intervortex spacing $a_0 \propto B^{-1/2}$, the interaction energy between different vortices in the same plane must increase with increasing B. However, the

Josephson coupling energy between two segments of the same vortex in adjacent planes—while strongly affected by the fixed γ of the superconductor structure—is shown to be independent of the magnetic field strength [Tink96]. Depending on the ratio of these two energies, one may expect for a system with given structural anisotropy either a three-dimensional character of the flux-line ensemble for prevailing Josephson coupling in small fields, or a two-dimensional structure consisting of pancake vortices in the limit of large fields and dominant in-plane vortex interaction. For $B \parallel c$ the crossover field B_{cr} separating these two regimes can be expressed as

$$B_{cr} \sim \Phi_0 \left(\frac{\lambda_{ab}}{s\lambda_c} \right)^2 \sim \frac{\Phi_0}{s^2\gamma^2} \tag{2.15}$$

where s is the interplanar distance [Vino90a, Fish91]. For large $B \gg B_{cr}$ the magnetic field dependence of the melting line $B_m(T)$ disappears because with increasing B the increasing stiffness of the vortex system just compensates for the decreased intervortex distance. In this regime the melting temperature quickly approaches a constant value $T_m(B > B_{cr}) \to T_m^{2D}$. If the magnetic field is reduced below B_{cr}, interplane vortex-vortex interaction becomes more important and the melting temperature increases leading to the qualitative dependence indicated in Fig. 2.3. For moderately anisotropic superconductors, such as YBa$_2$Cu$_3$O$_7$, a two-dimensional behavior has not been observed as the crossover field is at inaccessibly high fields $B_{cr} \gg 10$ T.[4] On the other hand for extremely anisotropic superconductors, such as Tl$_2$Ba$_2$CaCu$_2$O$_8$ with a mass anisotropy $\gamma \gg 100$ [Farr90], quasi two-dimensional responses have been reported even at low fields [Wen98]. But for superconductors with a γ on the order of 100, as found in BSCCO, a crossover with a predicted $T_m^{2D} \sim 30$–40 K could be observed [Tink96].

2.4 Disorder and Vortex Glass

Besides simply being able to immobilize a single vortex or ensembles thereof, randomly distributed pinning centers also introduce disorder into the vortex system. This quenched disorder influences a variety of phenomena related to vortex dynamics and has been shown to result in the formation of new phases of the vortex system which would not exist in a superconductor without defects. Larkin and Ovchinnikov (LO) first developed the concept of the destruction of the perfect flux-line lattice due to the presence of random point defects in their theory of collective pinning [Lark70, Lark79]. The pinning force on the FLL in such a system can be shown to increase as $N^{1/2}$, or $V^{1/2}$ if $N = nV$, where N is the total number of pinning sites in the volume V with a density n of pinning centers per unit volume. As the

---

[4] If the structural anisotropy of the YBa$_2$Cu$_3$O$_7$ system is artificially increased, as in superlattices or in ultra thin films, one can again observe such a crossover. However, this relates to sample structure rather than material properties of the superconducting system.

Lorentz force due to a transport current density J scales with JV it follows that the depinning current density J_{dp} at which the Lorentz force equals the maximum pinning force, will scale with $V^{-1/2}$ [Tink96]. Evidently, in the limit of large volumes a perfectly periodic and rigid FLL could not be effectively pinned by randomly distributed pinning centers because any position of the FLL relative to the super-conducting material would be equally favorable. A true superconducting phase without dissipation would thus not even exist for vanishing currents in a vortex solid at low temperatures.

2.4.1 Larkin and Ovchinnikov Model of Collective Pinning

Larkin and Ovchinnikov argued that the flux-line lattice represents an elastic medium and in a system containing disordered pinning it will deviate from the perfectly periodic and rigid arrangement. An increase in elastic energy due to a deformation of flux lines will compete with a gain in condensation energy due to the flux lines passing through more pinning sites. Therefore, the equilibrium vortex configuration will be a more or less distorted arrangement that minimizes the sum of both energies. The result, according to the LO theory of collective pinning, is an arrangement of the vortices that corresponds to the regular Abrikosov lattice within a given, limited correlation volume, whereas neighboring volumes will exhibit pinning-motivated shear and tilt distortions relative to each other. This situation is illustrated in Fig. 2.4, where the correlation volume $V_c = L_c R_c^2$ is indicated by its longitudinal and transverse dimensions L_c (along the direction of the magnetic field) and R_c, respectively. Ignoring uniaxial compression, the increase in elastic free

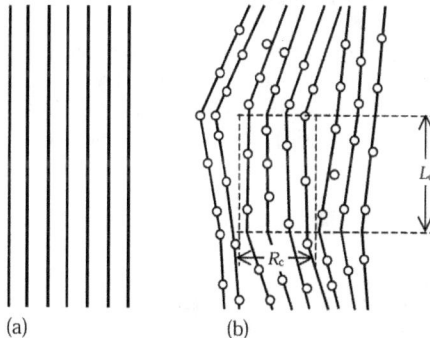

(a)   (b)

Figure 2.4: Schematic illustration of the coherence volume of the LO theory of collective pinning. (a) In the solid phase without pinning a periodic FLL parallel to the magnetic field exists. (b) The presence of random attractive pinning centers modulates the orientation of the FLL within each correlation volume V_c (defined by R_c and L_c) [Tink96].

17

energy per unit volume depends on the elastic moduli C_{66} and C_{44} for shear and tilt of the FLL, respectively, [Genn64] and for fractional distortions $s_s \sim \xi/R_c$ and $s_t \sim \xi/L_c$ has been given as $\frac{1}{2}\left(C_{66}s_s^2 + C_{44}s_t^2\right)$ [Tink96].[5] (Rough estimates of the moduli yield $C_{66} \sim \Phi_0 B/\lambda^2$ and $C_{44} \approx BH$.) Similarly, with the pinning force f acting over a length ξ the change in potential energy due to pinning inside the correlation volume can be approximated as $\sim \xi f N^{1/2}$ or $\xi f n^{1/2} V_c^{-1/2}$ per unit volume. The change in net free energy per unit volume resulting from distortion and pinning of the flux bundle then becomes

$$\delta F = \frac{1}{2}C_{66}\left(\frac{\xi}{R_c}\right)^2 + \frac{1}{2}C_{44}\left(\frac{\xi}{L_c}\right)^2 - f\xi\frac{n^{1/2}}{V_c^{1/2}} \tag{2.16}$$

Thus, one may obtain the dimensions of the resulting optimal correlation volume by minimizing (2.16) with respect to R_c and L_c and finds

$$L_c = \frac{2C_{44}C_{66}\xi^2}{nf^2} \qquad R_c = \frac{2^{1/2}C_{44}^{1/2}C_{66}^{1/2}\xi^2}{nf^2} \qquad V_c = \frac{4C_{44}^2 C_{66}^4 \xi^6}{n^3 f^6} \tag{2.17}$$

and hence the value of the net pinning energy corresponding to V_c [Tink96]

$$\delta F = -n^2 f^4 / \left(8C_{44}C_{66}^2 \xi^2\right) \tag{2.18}$$

This is easily understood in terms of smaller elastic energies and stronger or denser pinning sites resulting in decreased correlation volumes as a more frequent adjustment to the distribution of pinning sites becomes energetically favorable. A softer flux-line lattice therefore implies an increased depinning current density J_{dp}.

The implicit assumption of the Anderson-Kim model, that the activation energy U is a constant U_0 well below J_c according to $U \sim U_0(1 - J/J_c)$ cannot be sustained for the case of collective pinning by many weak pinning sites. For an equilibrium between the pinning force $\sim U_0/\xi$ and the Lorentz force $J\Phi_0 L_c$ on a segment one obtains a critical (depinning) current density $J_c \approx U_0/\Phi_0 L_c \xi$ below which flux-line motion will only occur due to thermal activation. Extending this argument to configurations of length scales longer than L_c and displacements larger than ξ a distinctive dependence of the activation energy U_0 on the current density J is found [Blat94, Tink96]

$$U(J) \sim U_0\left(\frac{J_c}{J}\right)^\mu \tag{2.19}$$

with a characteristic glass exponent $\mu \leq 1$. First found experimentally in flux-creep measurements of magnetization decay [Male90], this relation also leads to a characteristic nonohmic current-voltage dependence observable in transport measurements

---

[5] A more recent treatment of the shear moduli in collective pinning including the effects of anisotropy in HTSC's can be found in [Koga89, Sudb91].

$$V \propto \exp\left[-\frac{U_0}{kT}\left(\frac{J_c}{J}\right)^\mu\right]$$

(2.20)

2.4.2 Vortex Glass Theory of Fisher, Fisher, and Huse

While expecting the long-range order of the Abrikosov lattice to be destroyed in the presence of disordered pinning, no matter how weak, the theory of Larkin and Ovchinnikov nevertheless anticipates the highly nonlinear (glassy) response with vanishing linear resistance for small currents $J \ll J_{dp}$, as opposed to the theory of flux creep by Anderson and Kim with non-vanishing linear resistivity at all temperatures. Consequently, in systems containing disorder with the resultant loss of long-range FLL correlation, the question arises if there is a phase transition at a temperature (equivalent to the melting temperature) separating a vortex fluid phase with linear resistance from a glass (solid) phase with zero linear resistivity. M.P.A. Fisher in 1989 was the first to propose a vortex glass phase for systems with quenched disorder [Fish89a]. The extension to this model by Fisher, Fisher, and Huse (FFH) two years later [Fish91] describes in detail the properties of the associated states and the second order phase transition expected at the glass melting temperature T_g. Ever since, there have been a variety of modifications to this concept—relating, for example, to the effects of correlated disorder [Fish89b, Nels92, Nels93]—as well as numerous experimental reports of the observation of such glass transitions. The following treatment of the concept of vortex glass will be based on the theory of FFH as used throughout the literature, with special attention to the analysis of the transport measurements on $Bi_2Sr_2CaCu_2O_{8+\delta}$ and $YBa_2Cu_3O_7$ to be presented in Chaps. 3 and 4.

At the heart of the vortex glass theory lies a scaling argument. It considers a second order phase transition occurring at a glass temperature T_g which is characterized by the divergence of the correlation length ξ_{VG} and the relaxation time τ_{VG} of the vortex glass phase as

$$\xi_{VG} \propto \left|T - T_g\right|^{-\nu}$$

(2.21)

and

$$\tau_{VG} \propto \xi_{VG}^z$$

(2.22)

with static and dynamic critical exponents ν and z, respectively. If d describes the dimensionality of the system FFH argue that the electric field E should scale as $1/(\text{length}\times\text{time})$ and the current density J as $1/(\text{length})^{d-1}$ leading to the scaling hypothesis that two universal scaling functions $\mathcal{E}_\pm$ shall describe the dependence of E on J for temperatures above (+) and below (-) T_g as [Tink96]

$$\xi_{VG}^{z+1} E \approx \mathcal{E}_{\pm}\left(\xi_{VG}^{d-1} J\right) \tag{2.23}$$

In order to emphasize that J is scaled by a crossover current density $J_0 = kT/\Phi_0 \xi_{VG}^{d-1}$, which vanishes as $T \rightarrow T_g$, (2.23) is frequently written in the form [Koch89]

$$E \approx J \xi_{VG}^{d-2-z} \mathcal{E}_{\pm}\left(\xi_{VG}^{d-1} J\Phi_0/kT\right) \tag{2.23a}$$

The physical meaning of J_0 in both branches of the scaling functions becomes apparent if one investigates the properties to be expected in temperature ranges above and below the glass temperature [Fish91, Blat94].

In the vortex glass state the electric field is expected to assume an exponential dependence for low current densities

$$E(J) \sim \exp\left[-\left(J_c/J\right)^\mu\right] \qquad T < T_g \quad J < J_o^- \tag{2.24}$$

with a characteristic current density J_c and a glass exponent $\mu \leq 1$ [Fish91, Dekk92a]. Clearly, there is no contribution of linear resistance and E cannot be expressed in terms of a power law. In a double logarithmic plot of I–V isotherms from transport measurements this phase will exhibit a negative (downward) curvature as E decreases rapidly for $J \rightarrow 0$. If, in this low-temperature phase, J exceeds the crossover current

$$J_o^- \propto \left(T_g - T\right)^{\nu(d-1)} \tag{2.25}$$

the electric field is expected to approach asymptotically a power-law dependence $E \propto J^r$.

Precisely at T_g such a power-law dependence should persist over the entire current range resulting in a straight I–V line and visibly limiting the low-temperature glass phase of negative curvature. As the diverging length scale ξ_{VG} has to cancel out in (2.23) it follows that $\mathcal{E}_{\pm}(x) \sim x^s$ for $x \rightarrow \infty$, with $s = (z+1)/(d-1)$. Hence,

$$E(J) \sim J^{(z+1)/(d-1)} \qquad T = T_g \tag{2.26}$$

Similarly, in the fluid phase for $T > T_g$ one anticipates an asymptotic power-law behavior of the electric field above a crossover current density that vanishes on approaching T_g in the same way as (2.25)

$$J_o^+ \propto \left(T - T_g\right)^{\nu(d-1)} \tag{2.27}$$

However, below the crossover current the characteristics should change to a linear resistivity of a correlated vortex liquid phase

$$\rho(T) \propto \left(T - T_g\right)^{\nu(z-d+2)} \qquad T > T_g \quad J < J_o^+ \tag{2.28}$$

This ohmic behavior becomes visible in a double logarithmic plot of I-V curves as linear sections of the isotherms with a slope 1. Also, it is possible to identify this region in an analysis of a resistive transition $\rho(T)$ and extract the glass temperature T_g and the product $\nu(z-d+2)$ of the critical exponents (cf. Sec. 3.4).

Although it has not yet been mentioned in the literature, it is possible to express analytically the dependence of the electric-field strength E_0^+ on the crossover current density $J_0^+(T)$ for any given temperature in the fluid phase. Substituting (2.27) into (2.28) yields $\rho_0^+(T) \propto \left(J_0^+\right)^{(z-d+2)/(d-1)}$. As $E = \rho \cdot J$, multiplying both sides by J_0^+ one finds a power-law behavior for the crossover identical to that of the actual glass line

$$E_0^+\left(J_0^+\right) \propto \left(J_0^+\right)^{(z+1)/(d-1)} \tag{2.29}$$

The particular importance of this result will become obvious in Chap. 4, where it presents a possibility to extract, quite precisely, the dynamic exponent at temperatures above T_g from a clearly identifiable feature in the I-V characteristics.[6] While the analysis of resistive transition also relies on data taken at $T > T_g$ one only obtains the product of the dynamic exponents $\nu(z-1)$, which is generally plagued by a large error and from which z cannot be isolated. All other means of determining z, the identification of the I-V glass isotherm and a scaling analysis of all CVC's, include data at or below the glass temperature and do not depend on a such a clearly identifiable feature.

It is important to realize, that all of the above relations may break down at high temperatures or at high current densities due to the onset of free flux flow. Above the fluid temperature T_f the relation of (2.28) will go over into FFF with purely ohmic behavior over the entire current range due to higher thermal energy, less efficient pinning (because of the decrease in condensation energy and the divergence of ξ), and the temperature dependence of B_{c2}. This breakdown must result in a visible deviation in the analysis of resistive transitions as well as in the scaling analysis of I-V isotherms. While the CVC's in this range will remain ohmic, i.e. retain a slope of 1 in a double-logarithmic plot, the temperature dependence of (2.28) will not be valid [Deak93, Xeni93]. Also, even for temperatures $T < T_f$ free flux flow will set in at high current densities, because the Lorentz force will begin to strongly dominate the pinning force resulting effectively in free motion of the vortices. Excluding the latter region is of critical importance for a consistent analysis as this onset may result in a downward curvature at these high J for temperatures $T_g < T < T_f$, which could be mistaken for a sign of a glassy state.

---

[6] In the literature this dependence has rarely been scrutinized. Although the physical meaning of the crossover current density has been discussed, its specific appearance in the typical double logarithmic plot has not been treated. There exists a figure in the review of Blatter et al. where the location of the crossover for $T > T_g$ is indicated schematically and clearly appears to be *not* parallel to the glass line [Fig. 29 of Blat94]. However, already the resulting intersection of the crossover with the glass line is in contradiction with the premise that there is no ohmic region even at T_g even for lowest current densities.

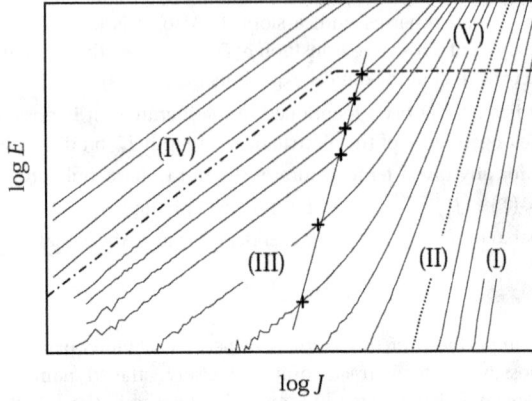

Figure 2.5: Typical I–V isotherms of a type II superconductor with disordered pinning for a fixed magnetic field. One can distinguish five different features: (I) the low-temperature vortex glass phase; (II) the I–V glass line at T_g; (III) the vortex fluid phase above T_g; (IV) the FFF region at temperatures $T > T_f$ for all current densities; (V) the region of the onset of FFF at high current densities and electric fields for all temperatures. Also indicated (+) are the locations of the crossover separating low- from high-current density behavior in (III). Note that the line connecting the crossover points in (III) is a parallel to the I–V glass line (II).

Figure 2.5 illustrates the typical appearance of a set of I–V isotherms taken at a constant magnetic field $\mu_0 H = 1$ T. The different regions are labeled as corresponding to (I) the low-temperature (solid) vortex glass phase with negative curvature; (II) the I–V glass line at T_g; (III) the high-temperature vortex fluid phase with positive curvature at the crossover to asymptotic behavior indicating the regime of the critical transition; (IV) the FFF region at temperatures $T > T_f$; (V) the onset of FFF at high current densities and electric fields for all temperatures. Also the crossover current density for $T > T_g$ has been indicated in region (III). According to (2.29) $E_0^+(J_0^+)$ appears as a parallel to the I–V glass line (II). The appearance of the CVC's can be related to the physical mechanism underlying vortex glass dynamics in terms of the divergence of ξ_{VG}, the length scale below which the vortex system is critical, and the inverse dependence of probing length and current density according to $\xi_{VG} \propto J^{1-d}$ [Blat94].[7] At temperatures above T_g a fluid response will exist at length scales $l > \xi_{VG}$ [visible as ohmic I–V curves (IV) at low current densities] whereas the system is in the critical state for $l < \xi_{VG}$ [characterized by the asymptotic behavior (III) of the isotherms at high J]. As the glass temperature is approached, the size of the critical region proliferates until the system is fully critical at T_g and a power-law dependence (II) prevails

---

[7] In the vortex glass model flux-line excitations are considered to occur in the form of vortex loops, whose characteristic size depends inversely on the transport current density as smaller excitations become energetically favorable with increasing current densities. Thus, at sufficiently small current densities in thin films, where the probing length will exceed the dimensions of the sample, finite size effects are expected to destroy the vortex glass transition.

at all current densities. Below T_g the flux-line ensemble is again considered to be critical for $l < \xi_{VG}$ (i.e. $J > J_0^-$) but exhibit a glassy response at larger length scales; these two regions cannot be distinguished easily due to their similar curvature.

The particular appeal of the vortex glass theory lies primarily in its ability to explain consistently the dynamic behavior of the flux-line ensemble within a single model over a wide range of temperatures, magnetic fields, and current densities. Alternative explanations—such as the variety of different pinning models including single vortex pinning, collective pinning, hopping of flux-line bundles (collective flux creep) [Feig89], depinning of the entire ensemble—always correspond only to very restricted ranges in T, B, and J. Even more enticing features of the VG model are the existence of a true superconducting phase (with zero linear resistivity) and the predicted universality of the character of the vortex glass transition. More specifically, the theory expects the critical exponents to be constant in all materials which exhibit this transition—an aspect which will be reexamined in Chap. 4.

2.5 B-T Phase Diagram of Layered HTSC's

In the previous sections the effects of thermal activation, flux-line interaction, dimensionality, and pinning on the dynamics of a vortex system have been summarized. The resulting B-T phase diagram for the vortex dynamics of an anisotropic high-temperature superconductor with quenched disorder is given in Fig. 2.6. For lowest temperatures the vortex ensemble exists in a solid state, which is expected to

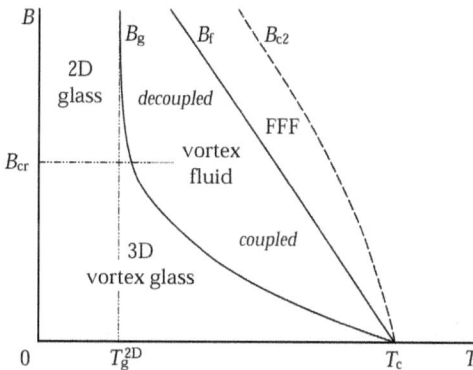

Figure 2.6:   Schematic phase diagram of a layered HTSC. In the presence of quenched disorder one distinguishes the glass (solid) and the fluid phases separated by the glass line $B_g(T)$ (rather than a melting line). At the fluid transition $B_f(T)$ the interacting fluid (critical regime of the glass transition or TAFF) changes into the FFF state of independent, free vortices. $B_{c1}(T)$ coincides with the $B \equiv 0$ axis in this representation. For weak coupling at high magnetic fields $B > B_{cr}$ the system may become two-dimensional with a constant glass temperature T_g^{2D} [Blat94, Tink96, Cohe97].

23

possess a glassy character in the presence of random point defects. The second order transition at the glass line $B_g(T)$ (equivalent to the first order melting transition of an Abrikosov lattice) separates this regime from the fluid phase at higher temperatures. For high temperatures and magnetic fields the flux-line interactions in the fluid become negligible and in the free flux flow state directly below B_{c2} one obtains free vortices. Independent of the solid or fluid nature of the vortex system, at high magnetic fields $B > B_{cr}$ the influence of interlayer coupling is strongly suppressed and a two-dimensional system results. While this crossover is of little consequence in the coupled or decoupled fluid, in the solid phase one observes a change of the substantial magnetic field dependence of the glass line in the three dimensional-regime to a constant glass temperature T_g^{2D} above B_{cr}.

In transport measurements it is possible to identify most of these lines, at least in principle, provided they appear at accessible temperatures and magnetic fields. The expected behavior of a vortex glass relating to T_g and T_f is well described theoretically allowing for a detailed analysis from CVC's and resistive transitions. However, the dimensionality crossover of the vortex system, evident in the phase diagram primarily at B_{cr}, is more difficult to identify as it corresponds to a gradual change rather than a sharp transition. In YBCO it is expected to occur at inaccessibly high magnetic fields ($B_{cr} \gg 10$ T) while for BSCCO, due to this compounds higher anisotropy, it should be observable ($B_{cr} \sim 1$ T). The crossover can be determined indirectly, for instance from the magnetic-field dependence of the activation energy or the suppression of the VG phase, which is predicted to exist strictly only in 3D systems. Alternative models for a quasi-2D or true 2D glass at high magnetic fields $B > B_{cr}$ have been suggested for a number of superconducting systems and will be considered in Chap. 3. As the dynamic response of a strongly pinned vortex ensemble (such as expected for low thermal energies due to the immobility of the flux lines) results only in a minute induced voltage, it will generally be difficult to investigate directly. Therefore, particularly in transport current measurements, the focus of an experiment will be placed on a fluid and the transition to the solid phase rather than on the solid phase itself.

3 Oxygen Stoichiometry and Vortex Dynamics in $Bi_2Sr_2CaCu_2O_{8+\delta}$

Due to their strongly layered crystal structure, which reduces the vortex coupling between the superconducting CuO_2 layers, $Bi_2Sr_2CaCu_2O_{8+\delta}$ compounds are excellent candidates for a detailed investigation of some properties of vortex dynamics in type II superconductors. Specifically, adjusting the dimensionality in the range between the limiting cases of 3D and 2D behavior by a controlled variation of the coupling strength may allow to investigate the models describing the dimensionality dependent properties of the vortex system. Most commonly, the mass anisotropy parameter γ is used as an indication for the degree of decoupling. For single crystals typical values on the order of $\gamma \sim 100$ have been reported [Farr89, Iye92, Mart92, Kota94], but an increase to 900 has been claimed for samples with strongly reduced oxygen content [Stei94]. In several Pd-doped samples even a change over four orders of magnitude in the resistivity anisotropy (ρ_c/ρ_{ab}) has been observed [Wink96]. This wide accessible range of γ as well as the possibility of over- and underdoping are particular advantages of BSCCO. In addition, it has been shown that the charge-carrier concentration and superconducting transition temperature appear to vary comparatively smoothly with oxygen annealing over a wide range [Mitz90], suggesting an equally gradual change of the coupling strength. Thus, BSCCO with a systematically varied oxygen content should allow significant *and* continuous tuning of the related superconducting properties, in particular adjusting the dimensionality from a near 3D to a quasi-2D behavior.

Previous investigations on the influence of oxygen doping have attributed a variation of γ to the influence of interstitial oxygen in the Bi_2O_2 double layers on the interlayer coupling by charge transfer effects. A recent study on the anisotropic resistivities of BSCCO single crystals with different oxygen content has shown the expected increase of ρ_c/ρ_{ab} with decreasing carrier concentration. However, a strong dependence of the anisotropy on temperature for any given sample was also observed [Chen98]. While the latter can be explained by the temperature dependence of the coherence length [Bale95], it complicates the analysis of the dimensionality of BSCCO samples solely by mass anisotropy: a change in dimensionality due to sample intrinsic (i.e. charge-transfer dependent) interlayer coupling may be obscured by the strong doping dependence of the transition temperature and the possible change in the transition width. Consequently, in an analysis of the anisotropy of three BSCCO films with varied oxygen content the critical current density J_c was investigated as a function of the reduced temperature $t = T/T_c$. This may allow to exclude the effects of a variation of T_c at least to some degree [Bale95]. An increase of J_c and, according to $\gamma = 1/J_c(0) = (1-t)/J_c(t)$, a decrease of γ with oxygen content was reported and related to improved pinning of vortices with stronger interlayer coupling. Though, a change of the density of pinning centers in the annealing process would also result in a change of J_c, leaving somewhat uncertain the extracted increase of relative values of the anisotropy by an order of magnitude (no absolute values were given).

Focussing on transport properties, this chapter will present an extensive comparative analysis of six $Bi_2Sr_2CaCu_2O_{8+\delta}$ films with almost identical as-grown properties and systematically varied oxygen stoichiometry. The first section will briefly deal with the process of sample preparation and characterization, focussing on a confirmation of the systematic variation of the oxygen content. The following sections will concentrate on the results of resistive measurements and current voltage characteristics. All experiments presented in this chapter utilized a geometry with the magnetic field parallel to the crystallographic c axis.[8] Of particular interest will be the extraction of activation energies and construction of a phase diagram, which seem most suitable for investigating the dimensionality of a group of samples as they allow to exclude to a large extent influences of other material properties that could obscure the effects of oxygen doping on interlayer coupling.

3.1 Sample Preparation and Characterization

3.1.1 Preparation and Annealing

The preparation and structural characterization of the samples as well as the Hall measurements were part of the diploma work of P. Haibach [Haib97]; only the key results shall be summarized in this section. The deposition method for the as-grown samples has been reported previously [Wagn91, Wagn94a]; a more detailed account of the entire procedure, in particular the method of oxygen annealing similar to that used by Balestrino et al. [Bale95], can be found in [Haib97, Voss99]. All $Bi_2Sr_2CaCu_2O_{8+\delta}$ films were produced by dc sputtering from stoichiometric targets onto [100] $SrTiO_3$ substrates. The deposition time of 120 min, resulting in a film

---

[8] Measurements with an angle $\theta = 0-90°$ between c axis and B, frequently used in an analysis of anisotropy, were also performed. However, the angular resolution of the apparatus could not be increased to a sufficient range ($\Delta\theta \ll 1°$) thus resulting in an experimental uncertainty too large for an insightful comparison of the samples.

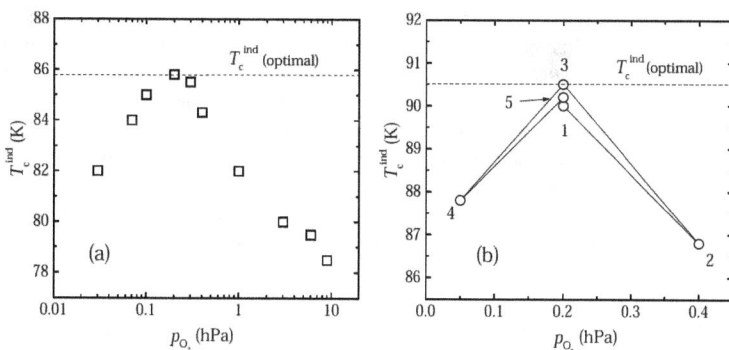

Figure 3.1: Variation of (inductive) transition temperature with oxygen annealing pressure and reversibility of the annealing process for two different test samples: (a) T_c exhibits a maximum for optimal oxygen content, whereas for lower and higher concentrations the transition temperature decreases. (b) For consecutive annealing steps (1–5) the transition temperature returns to its maximum value within < 1 K for the optimal O_2-pressure independent of the prior annealing state.

thickness of 4000 Å at a deposition rate of approximately 33 Å/min, was followed by a standard in-situ oxygen annealing step in order to yield samples initially with maximum T_c. As-grown samples were characterized by x-ray diffractometry and ac-susceptibility measurements of T_c. Depending on the intended increase or decrease of the oxygen content from the optimal value each sample was then subjected to an oxygen atmosphere of fixed pressure (10^{-3}–10^3 hPa) at a fixed temperature (500–600 °C) for 1 to 3 hrs in the final annealing process. In this step the concentration of interstitial oxygen bound in the Bi_2O_2 layers by Van-der-Waals forces is changed while the crystal's oxygen lattice itself remains unaltered. According to [Pres93] this results in a reduced T_c if the oxygen content is either increased or decreased from the optimal value (cf. Sec. 3.1.5). This dependence of the transition temperature on oxygen stoichiometry is clearly detected in the variation of T_c with the oxygen pressure p during annealing in consecutive annealing steps of one sample, shown in Fig. 3.1(a). While a maximum T_c was found for $p = 0.2$ hPa, it clearly decreases with lower as well as higher oxygen pressures.[9]

In order to investigate the reversibility of this annealing procedure, a different sample was consecutively annealed at optimal (1), increased (2), optimal (3), decreased (4) and finally once more at optimal (5) oxygen pressure with the transition temperature recorded after each step. Figure 3.1(b) indicates the return of T_c to the original optimal transition temperature of 90.3 K to within 0.4 K after each optimal annealing (1, 3, and 5). This allows to exclude the possibility of a degradation of sample quality in the annealing process.

---

[9] If the annealing time is too short the actual change in stoichiometry for any given set of parameters may depend strongly on sample properties, in particular surface morphology. A higher surface roughness may facilitate the diffusion of oxygen, which is presumed to occur primarily along the Bi_2O_2 layers rather than along the c axis.

27

For the experiments on transport properties ten BSCCO films were produced consecutively. To insure comparable material properties, such as homogeneity, defect density, and surface morphology, out of those ten films six samples of the highest quality and nearly identical inductive transition temperatures $T_c = 89 \pm 1$ K and transition widths $\Delta T_c \approx 1.7$–1.9 K were selected for further processing after the initial characterization. The annealing parameters, temperature and oxygen pressure, for these six samples of the annealing series are given in Tab. 3.1 along with their relative oxidation state. For ease of recognition the samples are generally identified by their name and a symbol indicating the oxygen stoichiometry, such as '(+)' for 'slightly increased' or '(--)' for 'decreased' content. For one sample of the original six BSCCO films, P90 (o), there is no data available from experiments at magnetic fields $\mu_0 H \neq 0$ as the film was destroyed in an early experimental stage during the zero-field measurements. Two years later a replacement sample, Z135 (o), with comparable properties was produced and the respective measurements could be repeated for a BSCCO film with optimal oxygen content. Nevertheless, since production parameters vary considerably with inevitable changes in the sputtering setup and procedure (e.g. due to adjustments to the heater or a new sputtering target) the new sample cannot be used as a direct substitute of the original. Material properties such as c-axis parameter or carrier concentration could not be considered conclusive evidence for the relative amount of oxygen in the sample and were not analyzed for the replacement Z135. However, for both samples P90 and Z135 the deposition process was optimized for a maximum T_c corresponding to an identical oxygen content (cf. Sec. 3.1.5) independent of other material properties, which supports the notion of comparable oxygen dependent interlayer coupling. This is further corroborated by the preliminary characterization (inductive transition) of both samples which exhibited similar electronic properties, in particular a transition temperature above 89 K and a narrow transition width. Including Z135 into the analysis as an optimal sample is therefore justifiable and will allow a systematic comparison of the vortex dynamics of all samples. Still, special attention to the possible influences of material properties (such as surface morphology and density of pinning centers) will be required to assure that observed deviations are actually due to the changed oxygen content instead of other material properties.

Table 3.1: Annealing parameters of the six BSCCO samples. p refers to the oxygen pressure and T to the sample temperature during the 3 hrs annealing process.

Sample	p (hPa)	T (°C)	Oxygen content	
P92	10^3	600	increased	(++)
P89	10^1	600	slightly increased	(+)
P90	2×10^{-1}	500	optimal	(o)
P88	7×10^{-2}	500	slightly decreased	(-)
P95	10^{-2}	500	decreased	(--)
P96	10^{-3}	500	strongly decreased	(---)

3.1.2 X-Ray Analysis

By x-ray diffractometry it was possible to determine the crystallographic c-axis parameter of all samples as described in detail in [Haib96]. It is found that c increases from 30.78 Å (++) to 30.87 Å (---) with decreasing oxygen content according to the oxygen annealing pressure, as displayed in Fig. 3.2. This dependence is attributed to the increase of excess oxygen incorporated between the Bi_2O_2 double layers consistent with previous reports [Mitz90, Emma92, Chen98]. Furthermore, it has been taken as an indication of a change in carrier concentration, which is in good agreement with the results of the Hall measurements confirming an increase of carrier density with oxygen annealing (see below). An overview of the oxygen stoichiometry dependent parameters including the results of the x-ray and Hall-effect measurements is presented in Tab. 3.2 (p. 34).

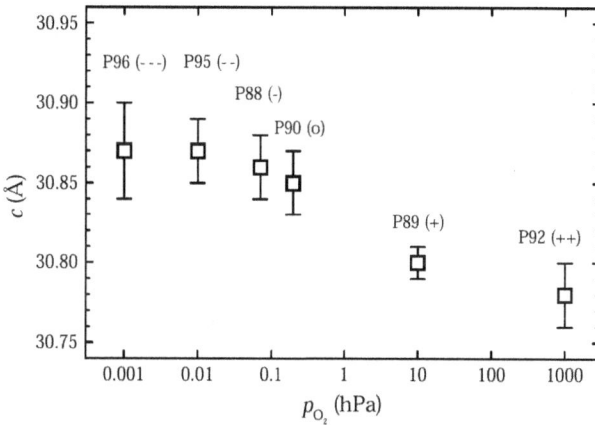

Figure 3.2: Variation of the c-axis parameter with oxygen pressure during annealing. The c-axis length decreases monotonously with increasing oxygen concentration.

3.1.3 Experimental Setup

After annealing and an initial characterization all selected samples were patterned into a geometry which allows standard four-point resistive measurements on a 1800 × 200 μm^2 and a 100 × 10 μm^2 measurement bridge as well as Hall measurements (Fig. 3.3). For all experiments presented in this chapter the larger stripline was used. Silver contact pads were evaporated onto the samples before the final annealing step to reduce contact resistance. All transport measurements were carried out in a ^{4}He-cryostat with exchange gas and temperature insulation chambers, which allowed to

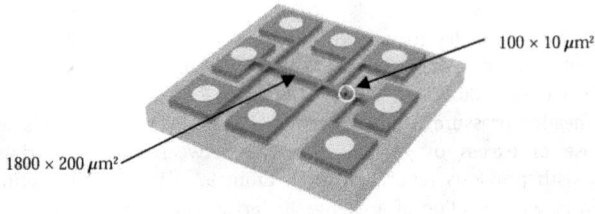

Figure 3.3: Structure of a patterned sample with two measurement bridges of $1800 \times 200\ \mu m^2$ and $100 \times 10\ \mu m^2$.

set a temperature between 4.2 K and room temperature with a stability better than 0.05 K. Magnetic fields up to 9 Tesla were applied parallel to the c axis of the samples and measured with a Hall probe incorporated into the sample cell. The transport current was supplied by a current source with a minimal setting and resolution of 5 nA and a maximum output of 0.1 A; square current pulses of 1 s duration and an interval time of 3 s between pulses were used. A nanovoltmeter recording the voltage drop over the sample at the end of each current pulse allowed the reduction of the noise level to approximately 20 nV.

3.1.4 Hall Effect

Hall measurements were performed in a magnetic field of 1 Tesla with a transport current of 0.5 mA (625 A/cm^2). Across the measurement bridge perpendicular to the direction of the current the transverse Hall voltage ρ_{xy} was recorded in a temperature range between 35 and 150 K. An increased oxygen content has been reported to contribute additional Hall-like charge carriers [Mitz90] and is therefore expected to result in a reduced Hall constant R_H, according to $n = 1/qR_H$, where n is the Hall carrier concentration and q the charge of a single carrier. While this relation is strictly valid only for a one band model, a systematic variation of n will give an indication of the relative oxygen content of the samples. A temperature dependence of the Hall constant is absent above the onset of the superconducting transition around 110 K, contrary to reports for YBCO. The absolute value of R_H decreases monotonously with increasing oxygen content as expected (Fig. 3.4).[10] Reasonable agreement is found with the results for BSCCO single crystals [Mitz90], where somewhat larger carrier densities (3–$4 \times 10^{21}\ cm^{-3}$) were determined. The qualitative behavior of the relative Hall-carrier concentrations from 1.3 to $3.0 \times 10^{21}\ cm^{-3}$ with increasing oxygen content is consistent with the analysis of the c-axis parameters as listed in Tab. 3.2 (p. 34).

---

[10] The type of anomalous behavior observable below 110 K for 4 of the samples has been scrutinized previously, e.g. in [Wang92]. Although not yet fully understood, it is negligible as far as the determination of the carrier concentration is concerned.

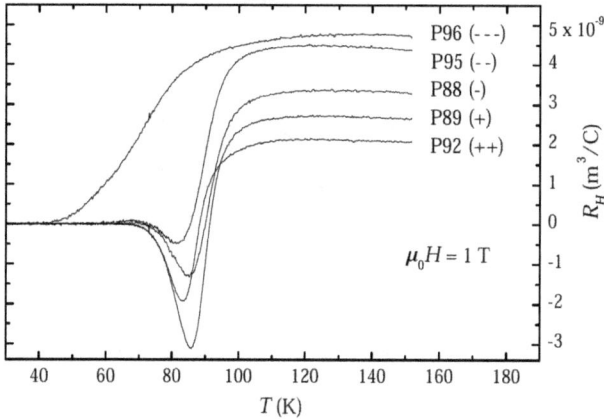

Figure 3.4:   Hall measurements of samples P92 (++), P89 (+), P88 (-), P95(--), and P96 (---). The Hall constant increases monotonously with decreasing oxygen concentration.

3.1.5 Aslamasov-Larkin Analysis

The critical temperature of the superconducting transition T_c is of special importance for this comparative study. First, it presents an additional option for estimating the respective oxygen stoichiometry of the samples. Furthermore, it is important to determine T_c in a consistent method with a small error for all samples as all subsequent analyses will be based on the reduced temperature $t = T/T_c$, where an inconsistently determined T_c for the various samples can strongly influence the conclusions. The zero-field resistive transitions are presented in Fig. 3.5 for all six samples of the original series and the additional sample. With the exception of the optimal sample, P90 (o), one finds a systematic decrease of the normal state resistivity ρ_n with oxygen content, which can be understood as an increase in charge-carrier concentration in agreement with the above results. It is evident that the transition temperature varies systematically yielding a maximum T_c for the optimal sample and decreasing values for over- and underdoping. The precise value of T_c was determined as the inflection point relating to the maximum slope of $\rho(T)$. Hence, $T_{c,ip}$ can be extracted easily and consistently from the resistive measurements of all samples. Use of the inflection point $T_{c,ip}$ instead of the downset temperature, for instance, evades problems arising from the relatively large transition widths $\Delta T_c \sim$ 10 K of BSCCO which further increase for over- and underdoped samples. The critical temperatures thus obtained are corroborated by the Aslamasov-Larkin (AL) analysis.

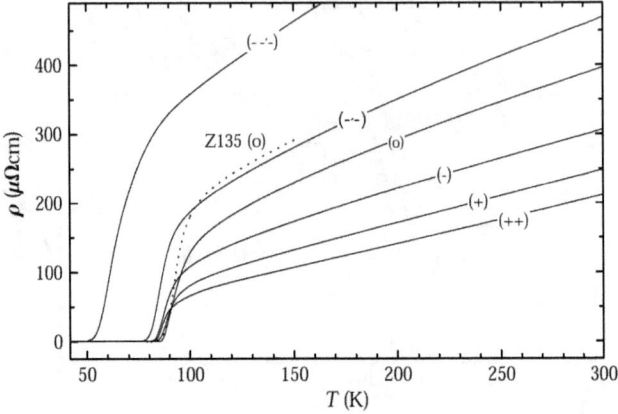

Figure 3.5: Resistive transitions of the six samples of the original oxygen annealing series and the replacement sample.

The comparatively wide transition regime in BSCCO has been attributed to the existence of superconducting fluctuations corresponding to regions in the sample with non-vanishing order parameter $\delta\psi$ well above T_c. As the density of super-conducting charge carriers is proportional to $|\delta\psi|^2$ these fluctuations will result in an excess conductivity $\Delta\sigma$, i.e. an onset of the reduction of resistivity for $T > T_c$ which was first described in detail by Aslamasov and Larkin [Asla68].[11] The excess conductivity of a sample is calculated from the resistive transition as $\Delta\sigma(T) = 1/\rho(T)-1/\rho_n(T)$, where the ρ_n is the extrapolation of the normal state resistivity determined by a linear fit to the high-temperature part ($T \geq 2T_c$) of the curve. The expressions for the Aslamasov-Larkin contributions of a 2D- and 3D-type superconductor are

$$\Delta\sigma_{AL}^{(2D)} = \frac{1}{16}\frac{e^2}{\hbar d}\left(\frac{T-T_c}{T_c}\right)^{-1} \qquad \Delta\sigma_{AL}^{(3D)} = \frac{1}{32}\frac{e^2}{\hbar\xi_0}\left(\frac{T-T_c}{T_c}\right)^{-1/2} \qquad (3.1)$$

where d is the thickness of a superconducting layer. Plotting $1/\Delta\sigma$ vs T one can extrapolate the AL transition temperature $T_{c,AL}$ as indicated in Fig. 3.6. In the cases of P90 (o), P88 (-), and P95 (--), $\rho(T)$ is well described by the two-dimensional AL model; however, for P89 (+) and P92 (++) [as well as for P96 (---) with a very broad transition] the linear section is more narrow and larger errors result. While for the overdoped samples the deviation from the $\Delta\sigma_{AL}^{(2D)}$ behavior may already indicate a transition to a more three-dimensional system due to increased interlayer coupling,

---

[11] This theory neglects the influence of Josephson coupling between consecutive layers. Although the resulting phase correlation of the order parameters can be explained according to Lawrence and Doniach [Law71], for the sole determination of T_c the Aslamasov-Larkin model is sufficient. For an extensive overview refer to [Skoc75].

nevertheless, for all samples the results of the Aslamasov-Larkin analysis fully confirm the values of $T_{c,ip}$ as listed in Tab. 3.2.

With the transition temperature it is possible to give a quantitative estimate of the samples' oxygen stoichiometry [Pres93]. From the amount of excess oxygen δ in the Bi_2O_2 double layers and the valence of Bismuth, $Bi^{3+p'}$, one may approximate the number of charge carriers per copper atom as $p = \delta - p' \approx \delta$ due to generally very small values of p'. Presland *et al.* have reported a universal parabolic relationship

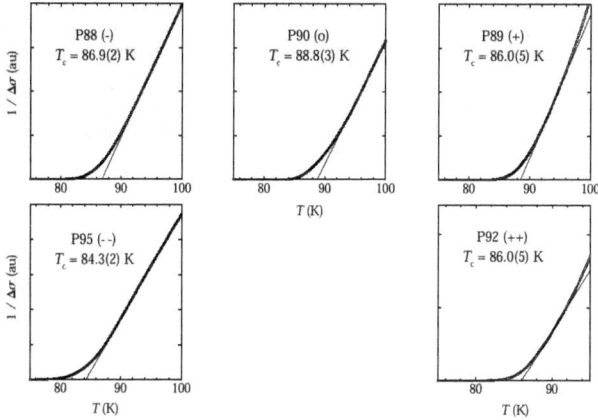

Figure 3.6: Aslamasov-Larkin plots of five samples for the determination of T_c. Straight lines indicate the extrapolation of T_c according to (3.1). The 2D model only describes the more two-dimensional systems of the optimal (o) and underdoped (-, --) films.

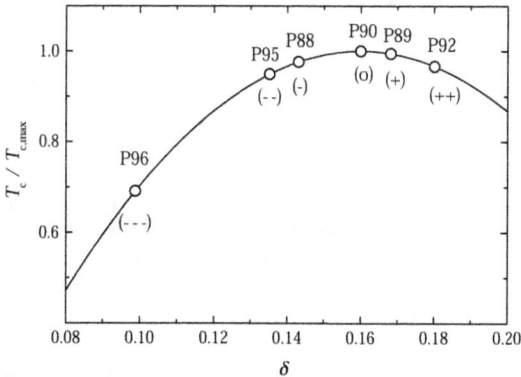

Figure 3.7: Oxygen concentration δ of the six samples as determined from the universal dependence of T_c on the doping level (see text for details).

33

between p and T_c for all p-type cuprate superconductors with $T_c(p) = T_{c,max} \cdot f(p)$ and $f(p) = 1 - 82.6\,(p - 0.16)^2$. Thus, a maximum T_c will be obtained for the optimal choice of $p = 0.16$. Samples are considered overdoped (oxygenated) in the range of $\delta \approx p > 0.16$ and underdoped (reduced) for $p < 0.16$. Identifying the transition temperature of the optimal sample with the maximum at $p = 0.16$ as shown in Fig. 3.7, one obtains the oxygen concentrations δ for all other samples according to their respective T_c, which qualitatively agree with the results of x-ray and Hall measurements.

Table 3.2: Comparison of oxygen stoichiometry effects in x-ray, Hall-effect, and T_c measurements for $Bi_2Sr_2CaCu_2O_{8+\delta}$ samples. c-axis parameter and carrier density n vary systematically with oxygen content. The critical transition temperatures of the Aslamasov-Larkin analysis $T_{c,AL}$ yield the same values as the inflection point of the resistive curves $T_{c,ip}$. From the latter the oxygen content δ of the annealed samples was extracted relative to the optimal sample, set as $\delta = 0.160$. For the replacement sample with optimal oxygen content, Z135, only $T_{c,ip}$ was extracted, which is sufficient for the subsequent determination of the reduced temperature $t = T/T_c$ needed in the comparative analysis of the vortex dynamics.

Sample		X-ray	Hall effect	Transition temperature		
		c-axis parameter (Å)	n (cm^{-3})	$T_{c,ip}$ (K)	$T_{c,AL}$ (K)	δ
P92	(++)	30.78(2)	$3.0(3) \times 10^{21}$	86.0(2)	86.0(9)	0.180(6)
P89	(+)	30.80(1)	$2.3(2) \times 10^{21}$	88.5(2)	88.5(6)	0.168(8)
P90	(o)	30.85(2)	–	89.0(2)	88.8(2)	0.160
P88	(-)	30.86(2)	$1.9(2) \times 10^{21}$	86.9(2)	87.0(1)	0.143(8)
P95	(--)	30.87(2)	$1.4(1) \times 10^{21}$	84.5(2)	84.3(2)	0.135(6)
P96	(---)	30.87(3)	$1.3(1) \times 10^{21}$	61.5(3)	62.0(9)	0.099(4)
Z135	(o)	–	–	92.5(2)	–	–

3.2 Activation Energy

For the analysis of the activation energy resistive transitions $\rho(T)$, with the voltage V across a sample recorded at constant current I and continuously decreasing temperature T, were performed at magnetic fields and temperatures ranging from 1 mT to 9 T and $2T_{c0}$ down to $T_{c,ds}(H)$, respectively. The broadening of the transition for increasing magnetic fields, evident in Fig. 3.8, has been widely reported for high-T_c superconductors (for a review see [Pals90] and references therein) and is generally understood in terms of thermally assisted flux flow [Tink88, Kes89, Kuce92]. Following the concept of the activation energy U_0 within the framework of the TAFF model (Sec. 2.3) the exponential form of the resistivity $\rho \propto \exp(-U_0/kT)$ (2.12) appears as a linear segment in the plot of $\ln\rho$ vs T^{-1} (Arrhenius behavior), with the value of U_0 depending directly on the slope. At high temperatures free flux flow will begin to set in, with the resistivity depending primarily on viscous drag. The regime of low

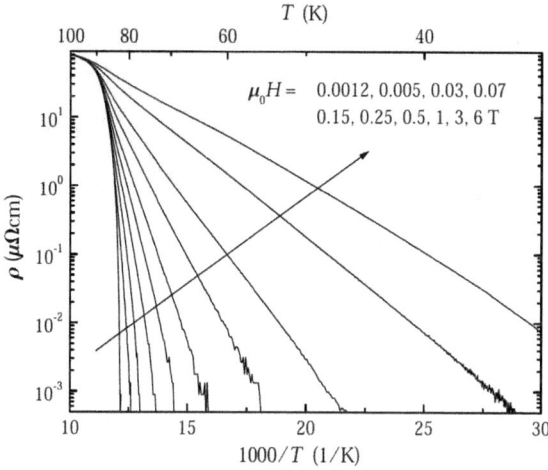

Figure 3.8: Resistive transitions of sample P89 (+) in magnetic fields $0.0012\,T \le \mu_0 H \le 6\,T$ plotted as $\log \rho$ vs T^{-1}. The broadening of the transition with increasing magnetic field can be explained within the TAFF model (see text). From the slope of the linear segments in this Arrhenius plot, the activation energy U_0 can be determined.

resistivity ($\rho < 10^{-2}\rho_n$), where TAFF behavior dominates, will be relevant for the extraction of U_0.

More interesting than its absolute value, however, is the magnetic-field dependence of the activation energy, which may reveal clues to the actual mechanism behind the TAFF behavior and is related to the dimensionality of the sample. Models that have been proposed and successfully applied to the interpretation of transport measurements include flux-lattice shearing (suggested in [Yesh88, Tink88]) and plastic lattice deformation (developed in [Gesh89, Vino90a, Vino90b]). Based on the theory of collective pinning and the activation of rows of fluxons, the former mechanism predicts a magnetic-field dependence

$$U_{sh} \propto B^{-1} \tag{3.2}$$

which has been observed repeatedly in the YBCO system [Pals89, Sun89, Hett89]. This inverse proportionality to the magnetic field arises from the cross sectional area of the flux-lattice unit cell $\sim a_0^2$ entering into the energy barrier via the volume of moved segments. The plastic deformation mechanism instead treats the creation of 'double-kink' excitations in flux lines, illustrated in Fig. 3.9. With the double-kink displacement l being limited by the average intervortex spacing $\sim a_0$ and the activation energy being proportional to the self energy of the vortex segments created in the ab plane, one obtains $U_{pl} \propto B^{-1/2}$, which has been observed in BSCCO films [Kuce92, Wagn94b]. In this system, the predominance of this mechanism has been

35

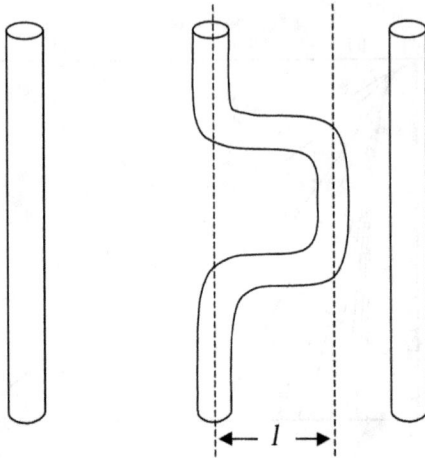

Figure 3.9: Schematic representation of the double-kink mechanism in plastic deformation.

attributed to the increased anisotropy and the reduced interlayer coupling. More specifically, calculating the activation energy from the anisotropic Ginzburg-Landau theory

$$U_{pl} = 2\epsilon_l a_0 \approx \frac{2\Phi_0^2}{4\pi\mu_0\lambda_{ab}\lambda_c}\ln\tilde{\kappa}\left(\frac{\Phi_0}{B}\right)^{1/2}$$ (3.3)

with the vortex energy per unit length ϵ_l, $\tilde{\kappa} = \sqrt{(\lambda_{ab}\lambda_c)/(\xi_{ab}\xi_c)}$, and $\lambda_c = \lambda_{ab}\gamma$ one finds

$$U_{pl} \propto B^{-1/2}$$ (3.3a)

$$U_{pl} \propto \gamma^{-1}$$ (3.3b)

The dependence of U_{pl} on B is simply a result of the change of the maximum average dislocation with magnetic field. Equation (3.3b) can be understood as a consequence of the increased in-plane mobility of adjacent vortex segments in neighboring layers (i.e. the decreased energy needed to create the vortex segments in the ab plane) in the case of increased anisotropy and weakened interlayer coupling. Thus, for YBCO with $\gamma \approx 5$–10 and a correspondingly high flux-line tension, plastic deformation requires much higher energies than lattice shear. However, for anisotropic materials such as Bismuth- and the Thallium-based cuprates the situation is reversed and plastic deformation dominates. Judging from the magnetic-field dependence of the activation energy found in this work for the oxygen-doped $Bi_2Sr_2CaCu_2O_{8+\delta}$ samples, actually both mechanisms seem to play a role in the vortex dynamics of this system, depending on the degree of anisotropy in a given sample.

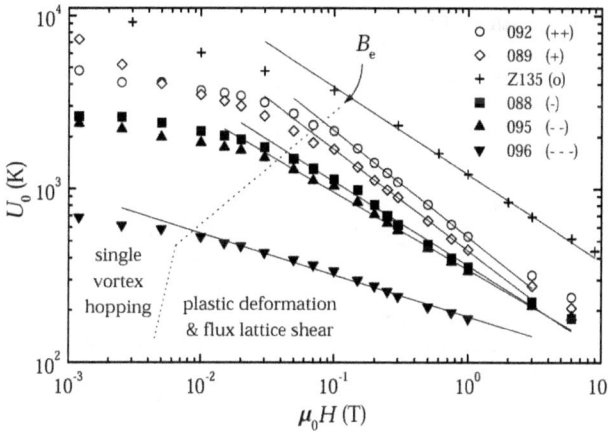

Figure 3.10: Activation energy U_0 over four orders of magnitude in magnetic field as extracted from the slope of the Arrhenius plots for six samples. For larger magnetic fields (above 0.01–0.1 T) all samples possess a regime with a power-law dependence of $U_0 \propto B^{-\alpha}$, but with different exponents α determined from the slope of the linear fits to this regime. These exponents show a systematic dependence on oxygen stoichiometry, as well as the absolute values of U_0 for the five samples of the original series. The visible crossover at low magnetic fields to a saturation regime $U_0 \sim$ const. also varies uniformly with the oxygen content of the samples.

U_0 was determined from the slope of the Arrhenius plot in the lowest decade of the resistivity accessible in all measurements, $0.1\ \mu\Omega\text{cm} \leq \rho \leq 1\ \mu\Omega\text{cm}$. While some curves did not display a linear section in the measurement range—primarily those of the overdoped samples at very low magnetic fields—most data closely followed the exponential TAFF behavior of the resistivity (2.12). The extracted values of the activation energy for all samples are plotted vs the magnetic field in Fig. 3.10. Several distinct features can be observed in this plot:

- The *absolute value of the activation energy* increases for all five samples of the original series monotonously with oxygen content regardless of magnetic field, as expected from (3.3b). A comparison with published results (e.g. 4–12 × 10⁴ K at 0.01 T and 2–6 × 10² K at 10 T [Kuce92], 5 × 10³ K at 0.1 T and 1.5 × 10³ K at 1 T [Wagn94]) reveals values well within the expected range but slightly smaller than previously found possibly resulting from different sample morphology.[12] At 0.1 T the minimum $U_0 \approx 300$ K for the sample with strongly decreased oxygen content, P96 (---); the maximum value, found in the overdoped sample, P92 (++), corresponds to 2000 K. A precise calculation of γ from this ratio according to (3.3) is

---

[12] The activation energy for the replacement sample, Z135 (o), also lies within the expected range but is clearly increased compared to the original series. As this sample was prepared under different circumstances and the preparation of BSCCO is known to be quite delicate, the changes in the absolute value of U_0 do not necessarily seem to indicate different activation *mechanisms* but rather a different effectiveness of the pinning centers. In fact, the annealing process itself may result in a change of the structure of the pinning centers.

not possible due to distinct magnetic-field dependencies for the various samples (see below) which lead to increasing ratios for lower B ($U_{0,P92}/U_{0,P96} \approx 3$ at 1 T, $U_{0,P92}/U_{0,P96} \approx 6$ at 0.1 T). However, using the relation $U_{pl} \approx 40000$ K $/ \gamma B^{1/2}$ [Goup97], one can estimate the anisotropy of the P89 (+) with $U_0 \approx 400$ K at $B = 1$ T to be on the order of 100, comparing well with results from B_e (see below) and previous reports. The relative anisotropies for 0.1 and 1.0 T are compiled in Tab. 3.3.

- For all six samples there exists a region in the *intermediate magnetic-field range* where the activation energy follows a power-law dependence $U_0 \propto B^{-\alpha}$. The exponents α determined from the slope of the linear fits to this regime yield values that increase systematically with oxygen stoichiometry from 0.24 (---) to 0.62 (++). This behavior is consistent even for the replacement sample Z135 (o), whose exponent $\alpha = 0.50$ agrees excellently with the values of the original series (see Tab. 3.3). Considering equations (3.2) and (3.3a), describing the expected magnetic-field dependence for flux-lattice shear and plastic flux deformation, respectively, this change in the exponent implies a transition from the more three-dimensional behavior ($\alpha \approx 1$, observed in YBCO) for samples with high oxygen content to a more two-dimensional behavior ($\alpha \approx 0.5$, observed in $Bi_2Sr_2CaCu_2O_{8+\delta}$ and $Tl_2Ba_2CaCu_2O_8$) for oxygen deficient samples. The intermediate values may be a result of both mechanisms contributing to dissipation but neither fully dominating the dynamics of the system. In terms of interlayer coupling this relates to a flux-line tension continuously increasing with oxygen content: for stronger coupling the activation energy of the double-kink mechanism rises and lattice shear becomes more favorable (higher α); for weakened coupling double kinks can be produced more easily and will dominate (lower α).

Table 3.3: Results of the analysis of the activation energy U_0. The exponent α of the power law was determined from the slope of the linear sections in Fig. 3.10. The crossover field B_e separates the regime of single vortex activation at $B < B_e$ from that of flux-lattice shear and plastic deformation at $B > B_e$. From B_e anisotropies for all samples are calculated according to (2.15). Following $U_0 \propto 1/\gamma$ relative anisotropies were also deduced from the absolute values of U_0 for the original series of samples (Z135 revealed a notably increased U_0). The relative anisotropies determined from B_e and U_0 show reasonable agreement, although the latter are somewhat field dependent due to varying exponents, i.e. different combinations of activation mechanisms.

Sample		Power law	Entanglement field			U_0	
		α	B_e (mT)	γ_e	$\gamma_{e,rel}$	$\gamma_{rel}(0.1T)$	$\gamma_{rel}(1.0T)$
P92	(++)	0.62(3)	90(10)	152	1	1	1
P89	(+)	0.59(3)	80(10)	161	1.06	1.27	1.19
Z135	(o)	0.50(6)	70(10)	172	1.13	–	–
P88	(-)	0.49(6)	40(5)	227	1.50	1.89	1.49
P95	(--)	0.45(4)	25(5)	288	1.90	2.08	1.58
P96	(---)	0.24(3)	7(2)	544	3.59	6.40	2.96

Yet, the question remains why α falls below the value of 0.5 predicted in Geshkenbein's double-kink model [Gesh89]. As it is derived by analogy from thermally activated motion (diffusion) of edge dislocations in atomic solids [Reed73], the model yields a dependence on $B^{-1/2}$ due to the lattice constant a_0 being considered as limiting the extension of double kinks. However, while the notion of a well defined lattice constant holds for the vortex *solid*, the vortex *liquid* with strongly varying intervortex distances can hardly be considered a lattice. Moreover, while the activation energy decreases as the vortex lattice becomes more dense, the increasing magnetic field leads to stronger vortex-vortex interactions and collective pinning effects may actually increase the activation energy compared to single vortex pinning thus leading to a more gradual reduction of U_0 with B in the double-kink mechanism. Previous reports on BSCCO do not contradict this hypothesis, as publications reporting exponents $\alpha \sim 0.5$ over extended magnetic-field ranges show clear, non-linear deviations (e.g. Fig. 2 of [Kuce92] and Fig. 7 of [Wagn94]) from this dependence and lower exponents have been found for BSCCO in thin films ($\alpha \sim 0.4$) [Kuce92] as well as in single crystals ($\alpha \sim 0.2$) [Pals88]. Likewise, the process of plastic deformation was reported as an activation mechanism at large magnetic fields in YBCO with a clearly higher exponent $a \approx 0.7$ [Abul96] giving further support to the thesis of interlayer coupling dependent exponents instead of constant $\alpha = 0.5$.

• At *low magnetic fields* the power-law behavior is cut off at some crossover field B_e and the activation energy seems to saturate below. While the transition is less pronounced in the underdoped films, this breakdown of the power-law behavior found at intermediate fields is clearly discernible for all samples and the value of B_e depends monotonously on the oxygen content (Tab. 3.3). Such a reduction of the activation energy below the $B^{-1/2}$ line has been reported for magnetic fields $\mu_0 H < 0.01$ T [Kuce92] and was explained as a crossover from collective to individual vortex pinning (which is approximately independent of B) due to the fact that the intervortex spacing a_0 may exceed the in-plane penetration depth λ_{ab}, the characteristic length scale for vortex interaction.[13] However, this approach cannot explain the clear variation of B_e from 0.01–0.1 T with oxygen content, observed in the present measurements. Furthermore, it does not account for the variable strength of flux-line tension: for a vortex with strong interlayer coupling plastic deformation will be more costly and single vortex hopping between adjacent pinning centers will already be more favorable compared to smaller double-kink displacements, i.e. at larger fields. Therefore, a *dimensionality induced* crossover from the double-kink mechanism, prevailing at larger magnetic fields $B > B_e$, to single vortex hopping, energetically advantageous at lower vortex densities $B < B_e$, seems the most suitable explanation for the doping dependence of the B_e (see below).

---

[13] In the presence of correlated disorder, such as columnar defects, a similar crossover to B-independent activation barriers was observed and explained by individual depinning processes [Goup97].

- One also observes for *high magnetic fields* $\mu_0 H \geq 3$ T a flattening of the slope of U_0 vs $\log B$. A crossover to the regime of 2D pancake vortices can be excluded as the activation energy in the pancake model is expected to follow the relation $U_0 \propto B^{-1}$ [Clem91]. Instead, a possible origin could be the small average intervortex distance ($a_0 \approx 25$ nm) which approaches the dimension of the flux-line core itself ($\xi \sim 2.5$ nm). If the displacements of the double-kinks cannot exceed considerably the size of the vortex structure itself this mechanism may become less effective for vortex activation and stronger collective effects can begin to dominate.

These observations can be interpreted consistently based on the concept of a modified vortex system dimensionality which depends on the strength of interlayer coupling:

The qualitative behavior of the activation energy, in particular its variation with magnetic field, is consistent in all samples. In the samples of the original series a dependence of the absolute value of the activation energy on the sample anisotropy was found in agreement with (3.3b) throughout the entire magnetic-field range. Furthermore, the variation of U_0 for these samples in the limit of single vortex hopping at low fields gives further support to the notion of interlayer coupling being influenced by charge-carrier density. Supposing pinning centers of uniform structure, i.e. active over a constant length l_p, one finds $U_0 \propto l_p \cdot \epsilon_1$ depending primarily on the self energy of the vortex. From (2.7a) with $\epsilon_1 \propto \lambda^{-2}$ and (2.2) with $\lambda^2 \propto n_s^{-1}$ one obtains $U_0 \propto n_s$, which is in qualitative agreement with the data.

Above B_e the vortex ensemble in all samples is in a state of reduced c-axis correlation allowing vortex deformations to occur. In this state a power law $U_0 \propto B^{-\alpha}$ results corresponding to the theoretical activation mechanisms of flux-lattice shear ($\alpha = 1$) and plastic deformation ($\alpha \approx \frac{1}{2}$). The actual value of the exponent differs for all samples depending on the relative contribution of the two mechanisms and it is clearly related to the samples' intrinsic anisotropy, with smaller anisotropies impeding the formation of double kinks and favoring lattice shear. Considering the exponents $\alpha = 0.25$–0.5 observed in the oxygen reduced samples, where the double-kink mechanism is expected to prevail, an exponent below the original theoretical prediction of $\alpha = \frac{1}{2}$ seems appropriate for pure plastic deformation. Below B_e flux-line activation appears to be dominated by single vortex hopping resulting in an increased correlation along the c axis and a vanishing dependence of U_0 on B.

In analogy to the crossover field B_{cr} separating the regimes of Josephson coupling and vortex interaction in the absence of pinning (cf. Sec. 2.3.2), the transition at B_e can be interpreted as a crossover from a disentangled vortex liquid at $B < B_e$ to an entangled liquid for $B > B_e$. (Both states are illustrated in Fig. 3.11.) However, the disentangled vortex liquid below B_e does not arise from an increased importance of interlayer coupling but instead from the vanishing of double-kink excitations, which, in terms of free energy, are too expensive at these large vortex distances. Nevertheless, a system of more rigid vortices results, i.e. a system of a more three-dimensional

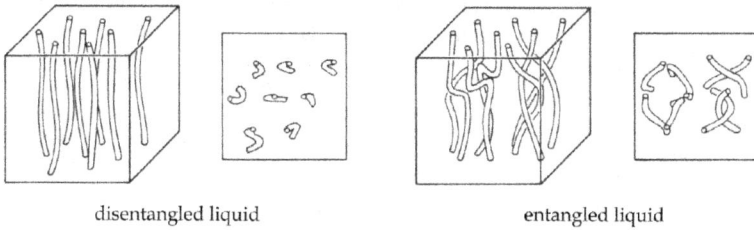

disentangled liquid   entangled liquid

Figure 3.11: Schematic representation of disentangled and entangled vortex ensembles [Blat94].

character, reflected in the vortex dynamics of the hopping of entire flux lines.[14] Although the transition at B_e is only gradual, the effective loss of long-range correlation along the c axis above B_e has been compared to the melting of the vortex lattice. As double-kink excitations become possible with smaller intervortex spacing above B_e, different segments along the same flux line will be able to move rather independently, corresponding to a more two-dimensional system.[15] Such a vortex ensemble under plastic deformation will thus resemble an entangled vortex liquid due to the double-kink excitations and corresponding vortex-vortex interactions and it has been noted that for an entangled vortex liquid the activation barrier should exhibit a power-law behavior [Blat94, Crab97]. The dependence of the specific value of B_e on the oxygen stoichiometry of the samples is a result of varying anisotropy: for stronger interlayer coupling the cost of plastic deformation is generally higher and the single vortex (3D) mechanism will already dominate at higher magnetic fields, while for weak coupling the double-kink excitations are favorable up to large displacements, i.e. down to smaller magnetic fields.

3.3 Critical Currents

Critical currents offer an additional approach to the scrutiny of the pinning process in superconductors, yet they provide less insight into the dimensionality and interlayer coupling than the activation energy as they relate to the pinning force rather than energy and thus are intrinsically related also to the density and structure of pinning sites.[16] Nevertheless, combined with the results of the analysis of the activation energy they support the interpretation of the influence of oxygen

---

[14] Also, the entanglement length, which depends on the intervortex spacing a_0 and thus increases for lower fields, may exceed the sample thickness preventing entanglement at low B.

[15] This should not be confused with the actual decoupling transition to independent 2D pancake vortices at B_{cr} [Kes90, Clem91].

[16] Higher activation energies will increase the critical current, as increased barriers heights will require a stronger driving force to be overcome. But if the extension of the barrier is widened considerably (i.e. the distance over which the driving force acts) the pinning force will be reduced even if the activation energy remains unchanged.

stoichiometry on sample anisotropy and flux-line tension. The critical currents were extracted from the I–V curves of all samples by a voltage criterion (10^{-6} V).

In Fig. 3.12 the temperature dependence of the critical current density for $\mu_0 H$ = 20 mT is displayed for samples P92 (++), P89 (+), P88 (-), and P95 (--). Since the critical transition temperatures vary considerably with oxygen doping the reduced temperature $t = T/T_c$ is used in this analysis. The absolute values of J_c compare reasonably to published values (10^4–10^5 A/cm^2) for the overdoped samples P92 and P89 [J_c(77 K) ~ 10^4 A/cm^2], whereas for the underdoped samples, P88 and P95, a considerably lower critical current density is found at all temperatures and magnetic fields. In fact, the critical current density appears to be directly related to the oxygen stoichiometry of the system: as dissipation sets in at lower driving forces for reduced oxygen contents the vortices are more mobile and less effectively pinned. This conforms nicely to the concept of weakened c-axis coupling as the lower flux-line tension will allow plastic deformation more easily and hence reduce the effectiveness of vortex pinning to existing defects. However, the reduction of J_c by several orders of magnitude within this sample series ($J_{c,P92} \approx 10^3 \cdot J_{c,P95}$ at t = 0.9) suggests that the structure of pinning sites may also be affected by the annealing process and contribute to the reduction of the critical currents.

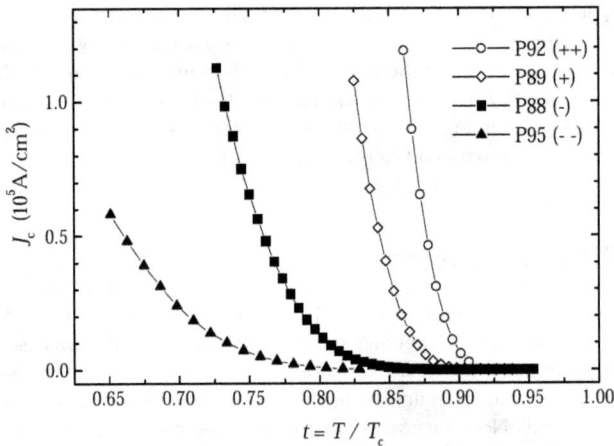

Figure 3.12: Critical current densities of four samples (++), (+), (-), and (--) as determined from I–V curves at $\mu_0 H$ = 20 mT by a voltage criterion (10^{-6} V).

3.4 Vortex Glass Phase

Strictly speaking, a vortex glass phase, according to the theory of FFH laid out in Sec. 2.3, should exist only in a three-dimensional ensemble of flux lines, whereas in a two-dimensional system the glass temperature is expected to vanish, $T_g^{2D} \equiv 0$, with the system only *approaching* a glassy state as T is lowered. However, the BSCCO samples investigated in this work, exhibiting neither truly two-dimensional properties (i.e. no decoupled pancake vortices) nor fully three-dimensional characteristics (except maybe at lowest magnetic fields in the overdoped samples), present an interesting intermediate system. The possibility of a system with TAFF behavior $\rho(T) \propto \exp(-U_0/kT)$ at $T > T_g$, such as Bi-2212, also possessing a glassy state for lower temperatures has been mentioned [Vino90b], a stack of weakly coupled 2D vortex glasses suggested [Fish91], and evidence for a vortex glass phase in the related $Bi_2Sr_2Ca_2Cu_3O_{10+\delta}$ system reported [Wagn95]. The following analysis will therefore concentrate on the influence of the varied flux-line tension on the character of a possible glass transition.

Typically, in transport measurements the transition from a fluid to a vortex glass phase is investigated by means of $I-V$ curves taken at temperatures above and below T_g for different magnetic fields. Distinguishing regimes of different curvature will, in principle, allow to identify the states of the vortex system at different temperatures and current densities, and a more elaborate analysis will yield the critical exponents and relevant temperatures of the glass transition.[17] The relevance of the analysis of CVC's is closely related to the measurement window as defined by the range of accessible electric fields and current densities. As T_g generally assumes low values in BSCCO the isotherms accessible in these transport measurements were located completely inside the fluid regime at all but the smallest magnetic fields, rendering an in-depth comparative analysis of the glass transition in all samples via CVC's impossible. The following analysis will concentrate on the Vogel-Fulcher relation for the resistive transitions.

According to (2.28), in the temperature region of the critical transition regime above T_g the resistivity will obey a power-law behavior $\rho(T) \propto (T - T_g)^{\nu(z-2+d)}$. More conveniently, this relationship can be expressed as

$$\left(\frac{d\ln\rho}{dT}\right)^{-1} = \frac{1}{\nu(z-d+2)} \cdot \left(T - T_g\right) \tag{3.4}$$

which allows to identify the relevant parameters of the glass transition in a corresponding plot: the product of the critical exponents from the slope of the linear section and the glass temperature from the extrapolated intercept as demonstrated in Fig. 3.13 [Safa92, Deak93, Xeni93]. Similarly, if one considers the expected TAFF behavior in such a plot, it follows from $\rho(T) \propto \exp(-U_0/kT)$

---

[17] A detailed description of the related concepts is given in Chap. 4.

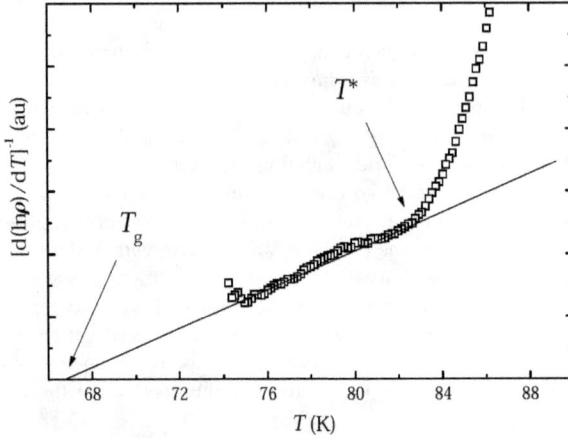

Figure 3.13: Example of a Vogel-Fulcher plot [sample P88 (-), $\mu_0H = 2.5$ mT] with the linear section indicating the regime of the critical glass transition. The slope of the linear fit yields the product $\nu(z-d+2)$ of the critical exponents and the temperature intercept corresponds to T_g. Also, the fluid temperature T_f, indicating the upper boundary of the critical regime and the onset of TAFF and FFF, may be determined from the deviation of the data from the linear behavior.

$$\left(\frac{d\ln\rho}{dT}\right)^{-1} \propto \frac{k}{U_0}\cdot T^2 \qquad (3.5)$$

Hence, the upper limit of the critical transition regime and onset of TAFF and FFF at T_f will be marked by a transition from the linear behavior of (3.4) to a parabola shape.

Resistive transitions of all samples in magnetic fields ranging from 1.2 mT to 9 T were thus analyzed and revealed two aspects which complicate the extraction of the parameters. (i) Owing to the magnetic-field dependence of the width of resistive transitions in HTSC's, the Vogel-Fulcher relation yields only very narrow linear sections at low B. Such a reduced fitting region naturally results in a comparatively large error of the slope and thus of the product of the critical exponents — an intrinsic problem of the Vogel-Fulcher analysis. Nevertheless, at the same time the uncertainty in T_g and T_f remains relatively small as both points are located very close to the data of the fitting region. (At larger B the fitting region extends over a larger temperature interval and the error of the slope is reduced.) (ii) Deviations from the expected line and parabola shape [according to (3.4) and (3.5)] become visible, in particular at higher magnetic fields ($\gtrsim 30$ mT). Typically, these plots display the expected linear section at low T and another linear segment of different slope at intermediate temperatures followed by the parabola-like increase at high T. This behavior is most pronounced for the overdoped samples but still visible in the slightly reduced sample P88 (-); for P95 (--) it disappears almost completely. The origin of this phenomenon is unclear. In light of the systematic change with magnetic field and oxygen stoichiometry it does not seem to be an experimental artifact. Possible causes for those deviations occurring at $T \lesssim T_c$ only in the overdoped samples at low and intermediate magnetic fields include a change of the entanglement length

$$l_z = \frac{C_{44}(T)}{kT} a_0^2 \approx \frac{C_{44}(T)}{kT} \frac{\Phi_0}{B} \tag{3.6}$$

which may exceed film thickness thus making plastic deformation unfavorable (especially the dependence of C_{44} on λ [Blat94] could explain the variation of the crossover temperature with oxygen content), as well as the divergence of the coherence length near T_c, which could lead to an increased c-axis correlation. However, with regard to the deviations at higher B and in the underdoped samples, a temperature induced decoupling of the flux lines along the c axis related to the temperature dependence of the entanglement field B_e (cf. Sec. 2.3.2) appears a more likely explanation. In this case, the increased thermal energy reduces the coherence between vortex segments of the same flux line in adjacent superconducting layers and could result in a changed temperature dependence of the resistivity. Recent results on BSCCO single crystals (e.g. investigations of the second magnetization peak [Khay96, Khay97, Deli97, for a review see Kes96], related theoretical analyses of the phase diagram of BSCCO [Baru98], and a non-monotonous dependence of the tilt modulus $C_{44}(B)$ with a maximum near $B \sim 1\text{--}10$ mT [Sude98]) suggest continued efforts in this direction, for example by pseudo-transformer experiments [Kosh94, Lope96a, Lope96b, Giam96]. In view of the influence of disorder in thin films, a further investigation of the magnetic-field dependence of the resistivity should be particularly interesting in this context.

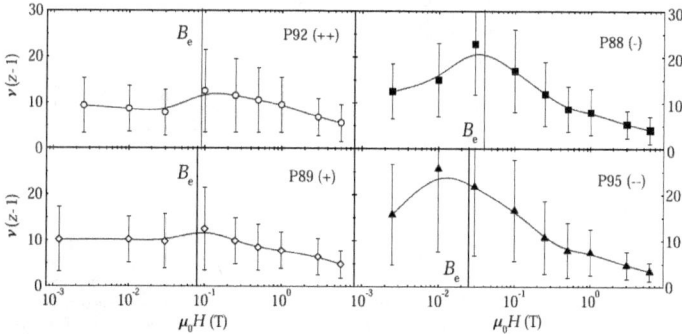

Figure 3.14: Product $\nu(z-1)$ of the critical exponents of the samples P92 (++), P89 (+), P88 (-), and P95 (--). Symbols indicate experimental data, solid lines are guides to the eye. The product is clearly magnetic-field dependent with a maximum at $\mu_0 H \leq 0.1$ T which coincides with the entanglement field B_e (indicated by a vertical line for each sample) as extracted from the activation energy.

The product of the critical exponents will be considered only briefly, since there exists no known relation to sample dimensionality. All results, showing considerable uncertainties, are compiled in Fig. 3.14. Despite the large errors a peak in the product $\nu(z-1)$ can be identified for each of the samples and is shifted to lower fields with decreasing oxygen content. There exists a remarkable coincidence between the magnetic field for which this peak is observed and the entanglement field B_e as determined from the activation energy. However, the vortex glass theory predicts critical exponents, which should be independent of magnetic field *and* sample. While increasing critical exponents have been observed at low magnetic fields in YBCO (see Chap. 4 and [Robe94, Noji96]) such a peak and subsequent decrease at even lower B cannot be explained within the VG model. A study of the low-temperature and magnetic-field regime by CVC's—allowing a much more precise (and separate) determination of ν and z—would be desirable.

Fortunately, in the Vogel-Fulcher analysis the resulting uncertainty for T_g and T_f is considerably smaller. With the extracted values the phase diagrams of Fig. 3.15 were constructed.[18] In all samples one discovers a constant value of $T_f \sim 0.9\ T_c$ for low $\mu_0 H$ ≤ 0.4 T followed by a reduction to $T_f \sim 0.7\ T_c$ at $\mu_0 H \sim 6$ T. While for samples with decreasing oxygen content T_f seems to shift to slightly lower temperatures and the

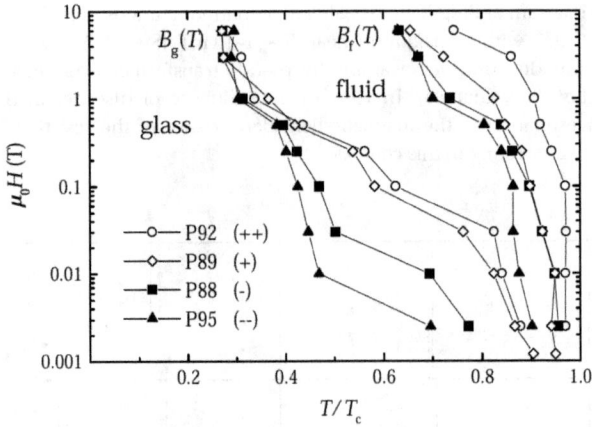

Figure 3.15: Logarithmic phase diagrams with glass and fluid lines, $B_g(T)$ and $B_e(T)$ respectively, separating the vortex glass phase at low T, the (fluid) regime of the critical glass transition for $T_g \leq T \leq T_f$, and the liquid phase with TAFF and FFF at high temperatures. The plateau in $B_g(T)$, appearing for all samples at the value of the crossover field B_e extracted from the activation energy, can be interpreted in terms of a crossover from a 3D vortex glass for $B < B_e$ to a weakly coupled stack of (2D) glass layers. An essentially decoupled 2D system appears to exits at high fields, $B > 1$ T, where the glass temperature is almost independent of B and the glass lines of all samples coincide.

---

[18] Apparently due to its extremely wide resistive transition, the glass and fluid lines are shifted to lower temperatures for the strongly underdoped sample P96 (---), which is therefore not considered in the quantitative analysis and omitted from the figures. Though, the qualitative behavior is preserved for this sample as well.

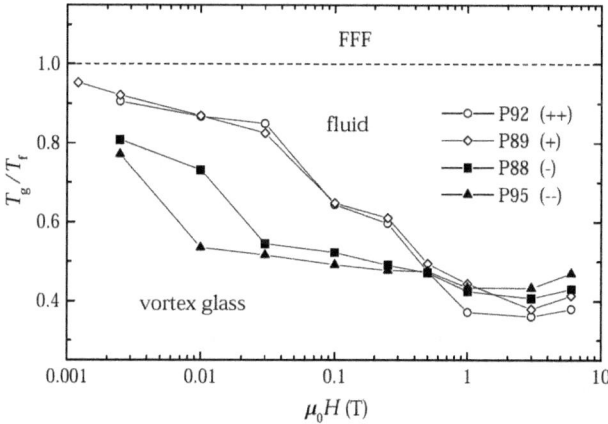

Figure 3.16: Relative width of glass and critical transition regimes T_g/T_f. Below 0.5 T the glass phase is suppressed in the underdoped samples.

onset of the reduction appears at somewhat lower fields, the phase boundary of the fluid regime appears largely identical in all films. For the glass line, however, the picture appears more complex. At high magnetic fields (>1 T) one observes nearly constant glass temperatures with increasing B coinciding at $T_g \sim 0.3\ T_c$ for $\mu_0 H \sim 6$ T in all samples. For low magnetic fields (<0.1 T) in the overdoped samples there exists another regime of almost constant $T_g \sim 0.85\ T_c$. The intermediate field range up to 1 T is characterized by a strong dependence of T_g on B appearing as a plateau in the glass line around 0.7 T_c. A similar plateau is also visible for the underdoped samples but occurring already at lower fields ~0.01 T. It seems reasonable to suspect that for these underdoped samples (as for the overdoped films) a region of magnetic-field independent high $T_g \sim 0.85\ T_c$ will also exist, yet at magnetic fields below the limits of this experiment. Due to the shift of the plateau for the different samples the critical region of the glass transition between T_g and T_f is widened considerably in low and intermediate magnetic fields for the oxygen reduced BSCCO systems, illustrated in Fig. 3.16 which plots the ratio T_g/T_f vs $\mu_0 H$. While this ratio is independent of sample stoichiometry above 0.5 T, below the ratio decreases monotonously with oxygen content.

The results of the Vogel-Fulcher analysis thus support the interpretation of the activation energy data presented above. The same crossover field B_e appearing in $U_0(B)$ is also observed in the magnetic-field dependence of the product of the critical exponents $v(z-1)$ as well as in the glass line $B_g(T)$. While a conclusive proof for a vortex glass is unavailable without CVC's taken at very low temperatures $T \ll T_c$, large current densities, and small electric fields, the resistive transitions strongly suggest such a glassy phase below T_g for $\mu_0 H < 1$ T. At higher magnetic fields all

47

samples may effectively become two-dimensional with a 2D melting transition at a constant T_g^{2D} above a crossover field B_{cr} (cf. Chap. 2.3.2, [Tink96]) agreeing well with the saturation of T_g above ~1 T and the coincidence of the glass line for all samples, regardless of anisotropy. Also, a 2D vortex glass, with $T_g^{2D} \equiv 0$ and no glass phase existing at finite temperatures (comparable to results for the Tl-2212 [Wen98]), cannot be excluded if one considers that the VF analysis considers only data above the glass phase. Moreover, FFH have suggested a deviation from the glass phase in BSCCO at magnetic fields on the order of 1 T [Fish91]. For low magnetic fields $B < B_e$ a high T_g and a narrow region of the critical glass transition is found, as would be expected for a more strongly three-dimensional system, such as 3D vortices of the single vortex hopping regime below B_e. The transition from disentangled vortices below B_e to an entangled system above B_e can also explain the steep decrease of $T_g(B \approx B_e)$ and the dependence of the value of B_e on oxygen content: as proposed by FFH, reduced but finite interlayer coupling at low temperatures may lead to a stack of weakly coupled vortex glass layers [Fish91]. For such a scenario one would expect a strong reduction in T_g at the crossover field and an anisotropy (i.e. coupling strength) dependence of the crossover field itself, as observed in all samples. The appearance of a wider critical transition region (cf. Fig. 3.16) may indicate the onset of such a layered vortex glass with reduced interlayer coupling. The hypothesis of a 3D glass at low B and a layered glass above B_e is in agreement with the observation that in the CVC's of all samples no glassy phase (i.e. negative curvature of the CVC's, cf. Chap. 4) appeared above B_e whereas for the overdoped samples (with the widest 3D vortex glass region) such a behavior was visible. Further support is also given by the plastic deformation mechanism indicating reduced layer coupling and by Bitter decoration experiments, which revealed a 3D system ~5 mT in BSCCO samples similar to YBCO [Gamm90].

3.5 Conclusions from Transport Measurements on BSCCO

The vortex dynamics of a series of six samples of Bi₂Sr₂CaCu₂O₈₊δ with systematically varied oxygen stoichiometry were investigated by transport measurements in a wide range of magnetic fields and temperatures. X-ray and Hall measurements confirmed the relative oxygen doping state of all samples as determined from the preparation process, and an Aslamasov-Larkin analysis of the resistive transitions made it possible to give an estimate of the actual value of δ for each sample according to the universal doping dependence of T_c. The observed monotonous variation of γ with oxygen content for all samples supports the influence of interstitial oxygen on interlayer coupling suggested in previous reports. In particular, the decrease of the critical current density on a normalized temperature scale indicates a less effective pinning for samples with lower δ and can be understood as a softening of the single flux line due to the weakened interaction between vortex segments in adjacent superconducting layers.

Moreover, a definite correlation between the oxygen concentration and the dimensionality of the vortex system was found. While the Aslamasov-Larkin analysis already indicated a deviation from the 2D model for the overdoped samples, this correlation is observed much more clearly in the activation energy and its magnetic-field dependence for the various samples. Specifically, it results in a change of the absolute value of U_0 in qualitative agreement with the theoretically predicted proportionality to γ^{-1} over the entire magnetic-field range for all samples of the original series. On the other hand, the characteristic change of the magnetic-field dependence in the intermediate field range can be explained by a variation of the contributions of the mechanisms of flux-lattice shear and double-kink plastic lattice deformation in the activation process as determined from the exponent α: for an increasing oxygen concentration $\alpha \to 1$ and the influence of flux-lattice shear continuously grows indicating a more three-dimensional character, whereas for the underdoped samples the double-kink process with a reduced exponent dominates corresponding to a more two-dimensional system.

The observed crossover of the activation mechanism at B_e to single vortex hopping with constant U_0 at low magnetic fields indicates a magnetic-field rather than sample dependent change in the dimensionality of the vortex dynamics. However, the

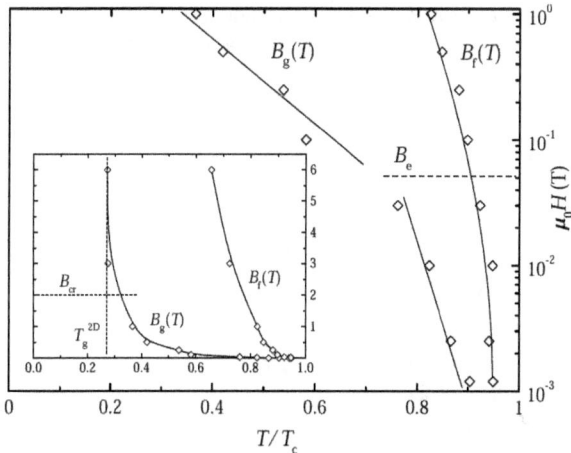

Figure 3.17: Experimental phase diagram for sample P89 (+) with glass line and fluid line. The entanglement field B_e is indicated in the double logarithmic plot. The crossover to a pancake system at B_{cr} and the 2D glass temperature T_g^{2D} become apparent for the linear scale of the inset. Symbols represent experimental data, solid lines are guides to the eye indicating the character of the phase boundaries. While the qualitative behavior appears largely identical for all other samples, the most decisive distinction is the strong variation of B_e with the oxygen content of the sample.

intrinsic sample dimensionality (i.e. the relative strength of interlayer coupling) shifts the value of B_e as lattice shear and plastic deformation already become energetically too costly for higher fields in more three-dimensional samples. This behavior is explained by a transition from an entangled vortex fluid at high to a disentangled fluid at low magnetic fields. The corresponding entanglement field B_e should depend on the flux-line tension as a higher rigidity of the vortices will make entanglement unfavorable in full agreement with the observed rise of B_e with oxygen content.

This notion is also supported by the results of the Vogel-Fulcher analysis of the vortex glass transition. In the obtained phase diagrams the behavior of the glass line changes at $B_g(T) = B_e$. At lower fields the glass transition occurs for all samples at $T/T_c \sim 0.8$–0.9. A disentangled vortex ensemble of a rather three-dimensional nature fits nicely to such a vortex glass phase with a relatively high T_g at low magnetic fields. Above the entanglement transition, however, due to the increasingly two-dimensional character of the system the glass temperature is strongly reduced and the creation of a vortex glass is suppressed in an intermediate field range $B_e < B \lesssim 1$ T, which coincides for each sample with the region of $U_0 \propto B^{-\alpha}$. At fields $B > B_{cr} \sim 1$ T, the glass temperature appears to saturate around $0.3\ T_c$, a value which agrees well with the expectations of the 2D glass (melting) temperature T_g^{2D} for a pancake vortex structure, as for example predicted in [Glaz91], with $B_{cr} \approx 1$ T and $T_m^{2D} \approx 30$–40 K, and reported in [Yama94] with $T_g^{2D} < 40$ K for $B > 0.5$ T. Combining all results for one sample one obtains a phase diagram of the vortex fluid and glass lines B_f and B_g, respectively, where the entanglement crossover field B_e, the 2D decoupling (pancake) crossover field B_{cr}, and the constant 2D (glass) melting temperature T_m^{2D} can be identified, as illustrated in Fig. 3.17 for sample P89 (+).

4 The Vortex Glass Phase of YBa$_2$Cu$_3$O$_7$

Suggested by Fisher, Fisher, and Huse (FFH) in 1989, the vortex glass phase was soon experimentally observed for YBa$_2$Cu$_3$O$_7$ in transport measurements on thin films by Koch et al. in 1989 [Koch89, Koch90] and on single crystals by Gammel et al. in 1991 [Gamm91]. Similar results have also been reported for other high- and low-temperature type II superconductors, e.g. Ta/Ge [Ruck97]. Though, while there has been strong evidence for the existence of a vortex glass phase there is still no clear agreement between theory and experiment, and the precise nature of a glassy state remains to be decided. In particular, the concept of an anisotropic glass scaling [Sawa98] and the magnetic-field dependence of critical exponents at very low magnetic fields [Robe94, Noji96] have been repeatedly considered. The origin of this variation in the universal exponents and the precise influence of finite size effects is as yet unclear. Furthermore, alternative causes for apparently glass like behavior have been offered. Flux creep has been claimed to result in similar transport characteristics at intermediate and high electric fields as a vortex glass [Copp90] and has since been mentioned as a possible explanation for deviations from glassy behavior at very low electric fields [Wen97]. The concept of flux creep, however, as proposed by Anderson and Kim in 1962, results in a fundamentally different state of the superconductor because the finite linear resistance at all current densities precludes the existence of a true superconducting phase with $\lim_{J \to 0} \rho = 0$ as it should exist in the vortex glass state.

In general, one of the few aspects concerning the investigation and interpretation of the low-temperature vortex phases of YBa$_2$Cu$_3$O$_7$ containing disordered pinning that is widely agreed upon in experimental as well as theoretical works is the necessity of extending the accessible voltage range in transport measurements to include very low electric fields [Copp90, Koch90, Blat94, Tink96, Cohe97]. Only with an extended measurement window including the low field range it will be possible to distinguish between the different models. The use of SQUID's for magnetization measurements with ranges extending from 10^{-6} to 10^{-11} V/m typically allows the investigation of the properties of the vortex glass phase at very low temperatures [Char95 and references therein]. This approach is less suited for the analysis of the transition of the system from a fluid to a vortex glass as the accessible current density range usually does not include a sufficient regime above and below the glass transition. Transport measurements, on the other hand, have typically been restricted to electric-field ranges of no more than 5 orders of magnitude extending only down to 10^{-5} V/m, so far. For the distinction between vortex glass and flux creep as well as for the unambiguous determination of critical exponents of a glass transition an extension of the accessible range to include high and low field data within the same measurement is indispensable.

In the beginning of this chapter the approach of extremely long measurement bridges as a solution to this problem of restricted measurement windows is presented briefly, describing in some detail the preparation and characterization of the samples. The chapter's central topic is the analysis of I-V isotherms covering a dynamic range exceeding eight orders of magnitude in electric field: different approaches revealed a second order phase transition well described by the general vortex glass theory as well as a magnetic-field dependence of the dynamic exponent z at low fields ($\mu_0 H < 0.3$ T). More importantly, a consequential sensitivity of the glass scaling to the measurement window and a dynamic exponent $z \approx 9$, clearly exceeding previously published results, were discovered. These findings are further corroborated by a Vogel-Fulcher analysis of resistive transitions. Alternative explanations other than a vortex glass are included in the discussion as well as prospects for the further investigation of the low-temperature (solid) vortex phase based on this new method.

4.1 Sample Preparation

The voltage sensitivity of a transport measurement setup depends primarily on the resolution of the voltmeter and the noise level of the entire circuitry. As the typical setup with a nanovoltmeter and coaxial shielded contact leads produces voltage sensitivities of about 10 nV there usually is very little room for significant improvement on the side of the apparatus. However, the *electric-field* sensitivity of a measurement will also be determined by the length of the measurement bridge. An increase of the typical length of the measurement bridge of 100 to 1000 μm to about 10 cm will immediately increase the electric-field sensitivity by several orders of magnitude from 10^{-4} to 10^{-5} V/m down to 10^{-7} V/m. While it is generally possible to create such a long bridge in the shape of a spiral on a substrate of 10 mm width and length, the obvious problem is the demand for samples of excellent quality, in particular of extremely high homogeneity over almost the entire substrate: compared to a 1000 μm bridge a 10 cm long spiral will require the density of such defects to be reduced by a factor of ~100. Likewise the process of patterning will be highly critical. Any speck of dust located on the sample during exposure, development, or etching can result either in an incomplete removal of the film between consecutive paths of the coil and a short circuit in the bridge or, even worse, in an excessive removal of film and a complete destruction of the measurement bridge.

For the work presented in this chapter, out of several available YBa$_2$Cu$_3$O$_7$ films prepared on SrTiO$_3$ two were chosen for patterning due to their small superconducting transition width as determined by inductance measurements on the unpatterned samples and due to their excellent crystallographic qualities as determined by x-ray analysis. Gold contact pads were evaporated onto the samples to minimize contact resistances and the measurement bridges were patterned by photolithography and wet chemical etching. As indicated above, extreme care and cleanliness had to be insured in this process to obtain intact measurement bridges. This included the removal and repeated application of photo resist to the sample if there appeared to be any defects (e.g. dust particles) in the intended area of the bridge in an applied layer of photo resist. The times of exposure, development, and etching also had to be determined very accurately: incomplete removal of sample material due to underexposure or insufficient etching could easily result in short circuits across the coil structure; excessive removal might render the entire sample useless in the case of a complete fracture in the bridge; finally, a moderate thinning would modify the current density and lead to increased Joule heating with the possibility of destroying the sample during measurement. Electric contacts were produced after patterning by ultrasonic bonding onto the gold contact pads. Both samples, Y121 and Y151 with a thickness of about 4000 Å, were patterned into identical 109 mm long and 50 μm wide unidirectional spirals with 9 coils of an average diameter of ~4 mm (Fig. 4.1). The structure allows measurements of the standard four-point method to avoid the influences of contact resistances.

Figure 4.1: Measurement bridge (unidirectional spiral, 109 mm × 50 μm) used for samples Y121 and Y151.

53

4.2 Sample Characterization

Rocking curve widths for the YBCO samples of this series were generally below 0.3° with excellent epitaxial c-axis oriented growth. The results of the ac magnetic-inductance measurements for samples Y121 and Y151 revealed transition widths of about 1 K. Yet, more important for evaluating the superconducting properties are transport measurements on the patterned film. As patterning procedures can seriously degrade sample quality due to chemical or thermal processes (without visible damage to the sample) measurements of the resistive transition in zero magnetic field can be considered the most solid way to establish the quality of the final sample.

Figure 4.2 presents the temperature dependence of the resistivity $\rho(T)$ for Y121 in zero (ambient) field in linear and logarithmic plots. The superconducting transition occurs at (midpoint) $T_{c,mp} = 90.0$ K with a transition width $\Delta T_c < 1$ K and does not display—neither in linear nor in logarithmic scale—any sign of a structure attributable to inhomogeneities or different phases. Usually, a transition width of less than 1 K with $T_c \approx 90$ K is regarded indicative of high quality YBa₂Cu₃O₇ films independent of the structure of the measurement bridge. If one considers the drop over six decades in the resistivity occurring along the entire 109 mm long bridge within 0.5 K, this narrow transition confirms the excellent quality of this sample. The data for sample Y151 with a $T_c = 90.5$ K and $\Delta T_c < 1$ K are comparable to that for Y121. Experimental results and analyses presented below are given for sample Y121 as the resistivity curve of Y151 displays a slightly wider transition. Still, the results for both samples agree well.

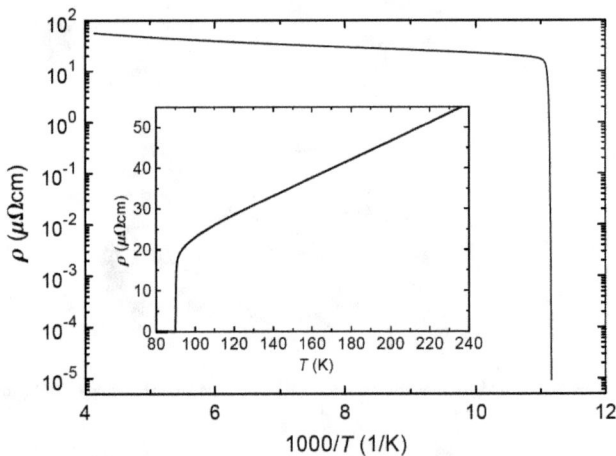

Figure 4.2: Resistive transition of the patterned sample Y121 in zero magnetic field.

The topological quality of the bridge structure is excellent in both samples with sharp edges, basically no underetching, and comparatively few defects. The latter appear to have originated from dust and other particles in or on the photo resist during patterning and have resulted in a slight widening of the bridge at some points. Due to the small diameter l_d of these defects ($l_d \ll 50\,\mu$m = width of the bridge) they do not obstruct transport experiments. A considerable reduction of the width, which could be of critical importance regardless of the length of the affected section, did not appear.

One should note that the kind of unidirectional coil used for this measurement bridge will always produce a cumulative magnetic self-field. Though, a rough estimate of the size of this effect by the Biot-Savart law yields $\mu_0 H_{center} = \Sigma(\mu_0 I / 2r_n) \sim$ 0.4 mT at the center of the 109 mm structure for currents up to 100 mA. The average self-field effects will thus be negligible in the range of external magnetic fields probed in this work ($\mu_0 H \geq 30$ mT).

4.3 Current-Voltage Characteristics

I-V curves are the most revealing kind of transport measurement in the study of vortex glass phase because they allow the extraction of the interesting parameters, specifically the glass temperature and the critical exponents, in several independent ways. Every single of these analyses may be obscured by different factors whose influences may be difficult to estimate or even detect, e.g. the onset of free flux flow at higher current densities and electric fields. Therefore, a sound investigation of the transition from the liquid to the solid vortex phase will strongly depend on careful consideration of such possible influences as well as on consistent results of the different approaches. The importance of combined analyses had already been recognized in the earliest studies of vortex glass—for example by Koch et al. who considered the behavior of the glass correlation length to check their results [Koch89]—yet, most reports have restricted themselves to the scrutiny of only one or at most two different approaches. The most common way of determining the glass temperature T_g and the dynamic exponent z is the identification of the I-V glass isotherm in CVC's. It will become obvious below that this method can yield misleading results, especially for z, as it is obscured by a restricted measurement window. With the deduced the relationship between electric field and current density at the crossover point from ohmic to asymptotic behavior for $T > T_g$ (2.29), a very revealing new approach, elaborated in Sec. 4.3.2, becomes possible. Furthermore, the glass correlation length will be considered and—as the most comprehensive examination—a scaling analysis of the glass transition will be conducted and the dependence of its quality on the measurement window considered.

4.3.1 *I–V* Glass Line

Determining the temperature of the vortex glass transition is most frequently done by identifying the only completely linear *I–V* isotherm in a double logarithmic plot of CVC's for a given magnetic field. This approach is based on the prediction of the critical behavior of $E(J)$ at T_g over the entire range of current density. Following (2.26) one finds

$$\log(E) \sim \frac{z+1}{d-1} \log(J) \qquad\qquad T = T_g \qquad\qquad (4.1)$$

Hence, in a double logarithmic plot this fully linear isotherm directly indicates the glass temperature and its slope, $s = (z+1)/(d-1)$, yields the dynamic exponent for the three-dimensional case $d = 3$, which can be safely assumed for YBCO. Yet, the prediction for this power law to hold at *all* current densities suggests that, in order to unambiguously detect the glass line, a considerable range in the current density and electric field should be desirable. If the line does not span several orders of magnitude in J and E, the observed 'linear' segment may actually just be part of the asymptotic branch of an isotherm above T_g whose upward curvature is compensated by the onset of flux flow.

Presented in Fig. 4.3 are the current-voltage characteristics of sample Y121 taken at $\mu_0 H = 1.0$ T for temperatures between 91.0 and 80.5 K with the electric-field range

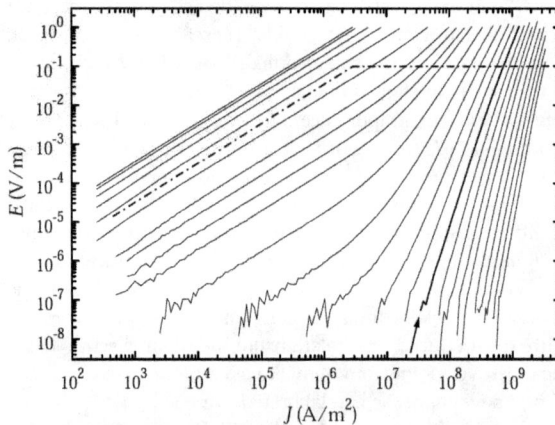

Figure 4.3: Double logarithmic plot of the CVC's for Y121 at $\mu_0 H = 1.0$ T and $T = 91.0$–80.5 K (from left to right) extending over an electric-field range from above 1 V/m down to below 10^{-8} V/m. Dashed lines denote the onset of free flux flow to be excluded from any analysis; the arrow indicates the resulting glass line at $T_g = 84.6$ K displaying a power-law behavior over the entire current range within the defined regime. Note that other isotherms at higher temperatures could appear linear if the electric-field window did not extend so far down.

extending from above 1 V/m down to below 10^{-8} V/m. The characteristic regimes of different vortex dynamics can be identified as described in Sec. 2.3.3. For temperatures above 85.8 K all isotherms display a linear section with a slope of 1 at low current densities and change to an asymptotic power-law dependence above the crossover current density J_0^*, resulting in a positive curvature indicative of the critical region above T_g. The I–V line at 84.6 K most closely follows a power-law over more than five orders of magnitude in the (low) electric-field range. Below this temperature all isotherms reveal a negative curvature as to be expected for a vortex glass state.

Upon inspection it is evident that some I–V curves (for $T > 84.6$ K) show both a positive curvature at low E and an negative curvature at high E. However, the latter is not related to the vortex glass state and can easily be explained as a result of the onset of free flux flow at high current densities and electric fields: the pinning energy becomes negligible compared to the Lorentz energy and the isotherms approach the ohmic appearance (i.e. slope of 1 as seen for $T \gtrsim 90$ K) of FFF at high T. Excluding this region ($E \geq 0.1$ V/m), as indicated by the dashed lines in Fig. 4.3, one then finds equation (4.1) fully satisfied only for the curve at $T_g = 84.6$ K, separating the regions of positive and negative curvature. The measured slope of this line, $s = (z+1)/(d-1) = 5.01$, yields a dynamic exponent $z = 9.0$. By the same procedure the I–V glass lines were determined for all CVC's, which exhibit similar qualitative behavior at all measured magnetic fields, 0.03 T $< \mu_0 H < 1.0$ T. The principal difference is the increase of the temperature interval of the critical region with magnetic field, resulting in a reduced uncertainty for T_g at lower fields. Obtained glass temperatures range from 89.5 K at 0.03 T to 84.6 K at 1.0 T; the slopes of the glass lines decrease from about 6.1 to 5.0, giving values for the dynamic exponent between $z \approx 11$ and $z \approx 9$. The complete data is compiled in Tab. 4.1.

An increase of the dynamic exponent at low magnetic fields has been reported previously [Robe94, Noji96] and will be discussed in some detail below. Though, comparing the recovered constant value of the dynamic exponent at higher fields ($z \approx 9$) with the results previously published for such transport measurements on YBCO ($z \approx 4$–6) [Koch89, Dekk92a, Ando92, Robe94, Noji96, Hou97, Sawa98] one finds a considerable increase which is difficult to explain, as the critical exponents are expected to be sample independent. Naturally, in light of these reports of lower z, the sole examination of the I–V glass line cannot be deemed conclusive evidence for an increased dynamic exponent and alternative methods of extracting z must be employed. One may already note, however, the evident influence of the electric-field window in the analysis of transport measurements: if the low field data were unavailable, a different isotherm might appear 'fully' linear—i.e. without positive curvature—thus resulting in the extraction of a higher T_g and a considerably lower slope and hence an underestimated z. This aspect will be further elaborated in Sec. 4.3.4.

Table 4.1:   Glass temperature T_g, fluid temperature T_f, dynamic exponent z, and static exponent v as extracted from the current voltage characteristics and resistive transitions of sample Y121 at various magnetic fields. CVC's were analyzed by means of the I-V glass line, the crossover current, and a glass scaling. For the resistive transitions a Vogel-Fulcher analysis was applied. Temperatures are given in Kelvin, magnetic fields in Tesla.

	I-V glass line		$E_0^+(J_0^+)$	Glass scaling analysis					Vogel-Fulcher analysis		
$\mu_0 H$	T_g	z	z	T_g	T_f	z	v	$v(z-1)$	T_g	T_f	$v(z-1)$
0.03	89.5(1)	11.1(8)	11.4(5)	89.50(10)	90.1 (2)	11.2(3)	1.25(5)	12.75	89.4(4)	90.0(3)	11.3(9)
0.10	88.6(1)	11.2(6)	10.0(4)	88.65(10)	89.8 (2)	10.4(3)	1.45(5)	13.63	88.8(5)	89.8(3)	13.8(9)
0.30	88.2(2)	9.0(5)	9.6(3)	88.15(20)	89.8 (2)	9.0(3)	1.60(5)	12.80	88.2(5)	89.6(3)	10.5(8)
0.60	86.4(2)	8.9(4)	8.7(3)	86.30(20)	89.2 (2)	8.9(3)	1.65(5)	13.04	86.6(5)	89.0(4)	11.7(8)
1.00	84.6(3)	9.0(4)	9.1(3)	84.55(20)	88.4 (3)	9.1(3)	1.65(5)	13.37	84.8(6)	88.3(4)	12.6(8)

4.3.2 Crossover Current Density

Although rarely scrutinized, the crossover current density assumes a unique position in the investigation of the vortex glass state in transport measurements as it represents a comparatively distinct feature easily identifiable in every single I-V curve in the temperature range above T_g. In the previous section the possible influence of the accessible electric-field range on a quantitative analysis has already been implied: the onset of free flux flow in the range of high electric fields may obscure the intrinsic curvature of the CVC's related to the glass transition and even the identification of an I-V glass line with a power law extending to very low electric fields leaves open the question if this dependence continues to still lower fields. This pertains to the extraction of the dynamic exponent from the I-V line's slope within the VG model as well as to general aspects of the VG theory, such as the distinction from the theory of flux creep, which predicts an ohmic regime at low current densities for *all* temperatures. Therefore, it is desirable to find a method which determines z independently of data at lowest electric fields. Identifying the crossover points in CVC's, i.e. the dependence of the crossover electric field E_0^+ on the crossover current density J_0^+ in accordance with (2.29), represents such a method.

The criterion used to define the crossover of a given isotherm is the increase of the resistivity by 10% above its linear (ohmic) value given by (2.28). For this purpose a straight line with slope 1 was fitted to the low current part of every curve with a linear section extending over at least one order of magnitude in current in the CVC's of Fig. 4.3. The point where the isotherms exceeded this linear fit by a factor of 1.1 yielded the crossover current density J_0^+ and electric field E_0^+. The choice of the factor of 1.1 was somewhat arbitrary—a different factor would simply have shifted all crossover points but it would not have changed the general behavior of $E_0^+(J_0^+)$.

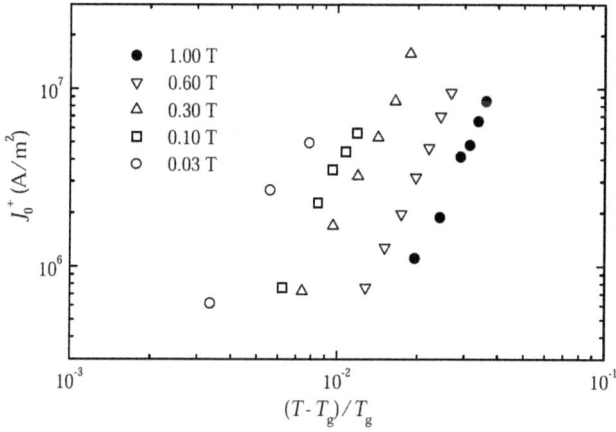

Figure 4.4: Double logarithmic plot of the crossover current density J_0^+ vs $T - T_g$ for magnetic fields $\mu_0 H = 0.03$–1.0 T. At low magnetic fields the critical regime of the glass transition is more narrow and fewer data points are obtained.

Choosing this small factor was possible due to a favorable signal to noise ratio even at low electric fields.[19]

In the first place the crossover current density is related to the glass correlation length ξ and thus to the static exponent ν. Combining $J_0^+ \propto 1/\xi_{VG}^{d-1}$ from (2.23) with the expression for ξ_{VG} from (2.21) for $T > T_g$ and $d = 3$ leads to

$$J_0^+ \propto \left(T - T_g\right)^{2\nu} \tag{4.2}$$

First suggested in [Koch89], this approach allows to check if the behavior of the crossover current is consistent over the entire range and to extract ν from the slope of (4.2) in a corresponding double logarithmic plot, presented in Fig. 4.4. All values of the static exponent extracted from J_0^+, given in Tab. 4.2, lie within the predictions of the VG theory and previous experimental findings ($\nu \approx 1$–2). The values suffer from the sensitivity of (4.2) to an uncertainty in T_g, which leads to a potentially large error. Especially at low fields $\mu_0 H \leq 0.1$ T the critical regime of the glass transition extends over such a small temperature range (less than 1 K for 0.03 T) that only few I–V lines with an identifiable crossover are available resulting in a particularly large uncertainty. Overall, however, the data conform well to the predicted power-law

---

[19] Defining the crossover according to its *slope*, as done in [Koch89] (10/9 of the linear value, i.e. a slope of 1.1), may actually obscure this analysis because the asymptotic branch of the isotherms in the critical region possesses a much larger slope for low temperatures $T \gtrsim T_g$ than at $T \lesssim T_c$. Hence, in the transition from linear to asymptotic behavior the slope will increase more slowly at higher temperatures, thus shifting the crossover point to higher current densities for these isotherms and resulting in a decreased z.

Table 4.2: Critical exponents and glass correlation lengths extracted from the analysis of the crossover current. The large relative error of the static exponent v for lower fields arises from the narrow temperature interval of the transition and the limited number of data points available from the CVC's. One observes that the glass correlation length ξ_{VG} clearly exceeds the average intervortex spacing a_0 of all investigated isotherms, even at low fields.

$\mu_0 H$ (T)	z	v	a_0 (nm)	$\xi_{VG,min}$ (nm)	$\xi_{VG,max}$ (nm)
0.03	11.4(5)	1.2(6)	263	348	983
0.10	10.0(4)	1.6(4)	144	327	885
0.30	9.6(3)	1.6(2)	83	194	904
0.60	8.7(3)	1.7(2)	59	249	876
1.00	9.1(3)	1.7(2)	45	182	720

behavior over the entire accessible range, which is confirmed in a plot of the glass correlation length according to (2.21) shown in Fig. 4.5. Using the extracted values of v one also obtains the predicted proportionality between ξ and $(T - T_g)^{-v}$. Moreover, the average intervortex spacing for the respective magnetic fields according to $a_0 \sim \sqrt{\Phi_0/B}$ is indicated in this plot (diamond symbols). $\xi > a_0$, considered a prerequisite for the possibility of vortex glass correlation effects, is clearly satisfied in all measurements.

Yet, the major importance of the crossover current density lies in its relation to the dynamic exponent. Identifying both J_0^+ and E_0^+ allows to extract z according to (2.29). The location of the crossover points in the individual isotherms, indicating the change from linear to asymptotic behavior, is denoted in Fig. 4.6 by solid dots. A fit to these points (dotted line) with the power law of (2.29) yields a slope of

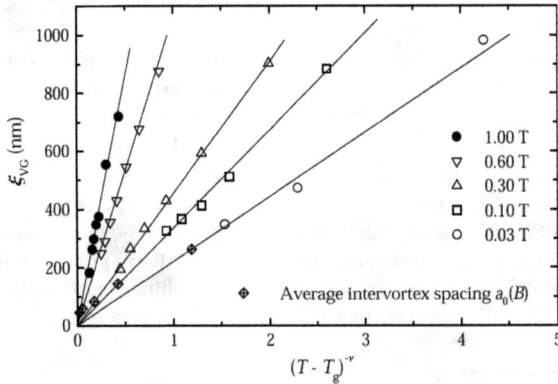

Figure 4.5: Temperature dependence of the vortex glass correlation length ξ_{VG}. The power-law behavior is preserved throughout the entire accessible temperature and magnetic-field regime. Diamond symbols indicate the average intervortex distances, which remain well below the minimum correlation length for each magnetic field.

$s = (z+1)/(d-1) = 5.04$ and thus a dynamic exponent of $z = 9.1$ for $\mu_0 H = 1.0\,\text{T}$ in excellent agreement with the values extracted from the I–V glass line ($s = 5.01$, $z = 9.0$). For all magnetic fields the recovered results as summarized in Tab. 4.2 (and compared with the alternative analyses in Tab. 4.1) support precisely the findings of the I–V glass line analysis. The significance of this crossover analysis arises from the fact that there is a consistent dependence of E_0^+ on J_0^+, which is identifiable at high electric fields and not influenced by the onset of free flux flow as in the case of the I–V glass line, for example. The uncertainty as to what happens at yet lower electric fields does not affect the extracted value of z. A change of the power-law dependence in that range would rather refute the vortex glass model altogether. Thus, the conclusion to be drawn is twofold: (*i*) Obviously, the agreement with the predictions of the VG model is very good, yielding precisely the predicted power-law dependence over the entire measurement window. Although this fact is no unambiguous proof for the existence of vortex glass it clearly endorses this model as opposed to alternative theories which cannot explain the observed behavior, such as flux creep. (*ii*) Within the VG model the data rule out the possible influences of finite size effects (discussed below), give very strong support to the notion of large dynamic exponents $z \approx 9$ and clearly reject the previously reported experimental values $z \approx 4$–6 as well as the theoretical estimates $z \approx 4$–7.

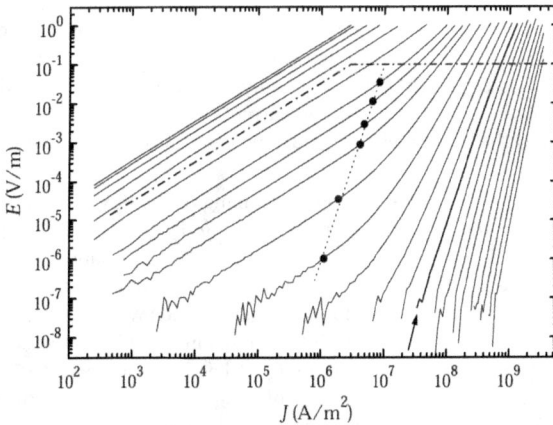

Figure 4.6: Double logarithmic plot of the same data as in Fig. 4.3 ($\mu_0 H = 1.0\,\text{T}$) with the location of the crossover points indicated by solid dots for isotherms displaying linear (ohmic) behavior at low J and asymptotic behavior at high J. The extracted dependence of the crossover field on the crossover current density follows very closely the relationship of (2.29). The dashed line gives the best fit to this power law with a slope of $s = (z+1)/(d-1) = 5.04$ corresponding to $z = 9.1$ in excellent agreement with the results of the I–V glass line. (Note the identical slopes of this linear fit and the I–V glass line indicated by the arrow.)

4.3.3 Vortex Glass Scaling

Using the ansatz of (2.23) one can determine T_g, T_f, z, and ν simultaneously for a given set of isotherms at a fixed magnetic field via a scaling analysis of the CVC's. According to (2.21) and (2.23) for a glass transition all I-V curves within a scaling regime can collapse onto the two common branches of the scaling function $\mathcal{E}_\pm$ above (+) and below (-) T_g if the scaled quantities

$$(E/J)_{sc} \equiv (E/J) \cdot |T - T_g|^{\nu(d-2-z)} \tag{4.3}$$

$$J_{sc} \equiv (J/T) \cdot |T - T_g|^{-\nu(d-1)} \tag{4.4}$$

are used. The scaling regime will be limited by the onset of free flux flow at high temperatures and high electric fields as discussed in Sec. 2.3.3 and indicated by the dashed lines in Fig. 4.3. In particular, above the fluid temperature T_f the system consists of non-interacting free vortices, the temperature dependence of the resistivity (2.28) will no longer be valid and scaling not possible . However, for an appropriate choice of T_f, T_g, z, and ν a complete collapse will be obtained. This method has been employed repeatedly with varying degrees of success indicating that it is to be applied with care [Koch90, Ando92, Robe94, Wölt95, Noji96, Hou97, Ruck97, Sawa98]. It was demonstrated that a scaling of reasonable quality does not necessarily imply the existence of a vortex glass transition: Coppersmith et al. even achieved a fair scaling with theoretical curves based on the flux creep model [Copp90]. Koch et al. subsequently showed that their experimental data of the YBCO system allowed a scaling of superior quality [Koch90]. It is evident that—besides additional analyses, e.g. of the crossover current—an excellent scaling is necessary in order to support the VG model and extract meaningful critical exponents.

As there is no mathematical criterion describing the quality of a scaling for a given set of parameters the procedure for determining T_g, z, and ν relies primarily on visual inspection. Usually, one will begin with a first guess and—judging from the deviations visible in the subsequent scaling—consecutively adjust the parameter values. By this method the CVC's for all magnetic fields were analyzed. As an example, the resulting scaling for $\mu_0H = 1.0$ T is shown in Fig. 4.7. This scaling exhibits an excellent collapse without deviations for $T_g = 84.55$ K, $z = 9.1$, and $\nu = 1.65$, with each curve extending over up to 7 orders of magnitude in electric field. The fluid temperature $T_f = 88.4$ K is then determined as the upper limit of the scaling regime above which this optimal scaling breaks down (as expected within the VG model).

For all magnetic fields the achieved scalings are of comparable quality and the obtained parameters coincide precisely with those from the I-V glass line and the crossover current analysis (Tab. 4.1) including the increased dynamic exponent $z \approx 9$. In no instance could a scaling of comparable quality for the full data range be achieved for any other set of parameters (outside the given experimental error),

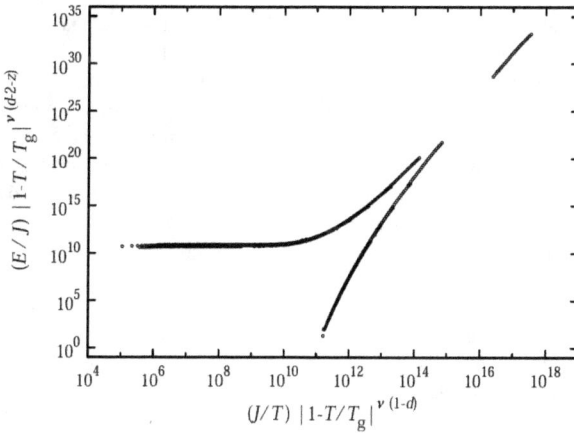

Figure 4.7: Glass scaling of the data from Fig. 4.6 with (free) scaling parameters T_g = 84.55 K, z = 9.1, and ν = 1.65. The region of the onset of free flux flow was excluded with T_f = 88.4 K. Excellent scaling is achieved for the entire accessible electric field ranging from 10^{-1} V/m down to below 10^{-8} V/m.

regardless of how many of the parameters were changed at a time. However, an artificial reduction of the accessible electric-field window, excluding low field data, did lead to a broadening of the range of suitable parameters giving a satisfactory scaling.

4.3.4 Influence of the Measurement Window

Although the analyses of the I–V glass line, the crossover current, and the glass scaling all produce consistent results and unambiguously support the notion of a dynamic exponent $z \approx 9$, the question remains, why previous transport measurements relying on CVC's from more restricted electric-field windows have almost consistently produced values of $z \approx$ 4–6 [Koch89, Ando92, Robe94, Wölt95, Noji96, Hou97, Sawa98].[20] These reports were based on analyses of the I–V glass lines and glass scalings of CVC's similar to this work. However, they did not utilize the power-law behavior of the crossover current in order to confirm the dynamic exponent, and the sensitivity of the measurements was confined to relatively small windows of high electric fields. The latter typically extended over four [Koch89, Ando92, Robe94, Noji96] to six [Hou97, Sawa98] orders of magnitude and were limited to electric fields in excess of 10^{-5} V/m.

---

[20] Increased values of z have been reported only for low magnetic fields or due to finite size effects. This aspect will be discussed below.

In the discussion of the I-V glass line the possible effects of a restricted measurement window have already been implied: the low electric-field data is crucial to the identification of a fully linear I-V curve, as opposed to the linear *segment* in the upper asymptotic regime present in *all* I-V curves of the critical regime of the glass transition (for $T \gtrsim T_g$). This explanation for underestimated dynamic exponents extracted from I-V glass lines becomes all the more likely, considering the observed onset of FFF above approximately 0.1 V/m in the samples of this work (cf. Fig. 4.3) and the electric-field ranges exceeding 1 or even 10 V/m [Hou97, Sawa98] used in previous experiments. Even allowing for different sample qualities, for example regarding activation energy and pinning site density, a difference of two to three orders of magnitude in electric field characterizing the onset of FFF seems unlikely. Also, one notes that the FFF features of CVC's taken with measurement windows limited to high electric fields (coincidentally) also seem to follow the predicted characteristics of a VG transition. If one neglects the regime below 10^{-5} V/m in Fig. 4.3 the remaining CVC's appear quite similar with the I-V glass line simply shifted to the left, i.e. to a higher temperature, and—due to this 'new' glass line's lower slope—with a smaller z. However, this is the result of the onset of FFF at high J even for $T > T_g$, which leads to a negative curvature just as expected of a vortex glass. This overall similarity of high electric-field free flux flow CVC's to the features of the actual glass transition seems to be one of the causes of the low dynamic exponents extracted from glass scalings of these data.

Artificially restricting the range of the measurement data to a higher electric-field window (such as 10^{-1} V/m $\geq E \geq$ 10^{-5} V/m, which is still large and extends to low

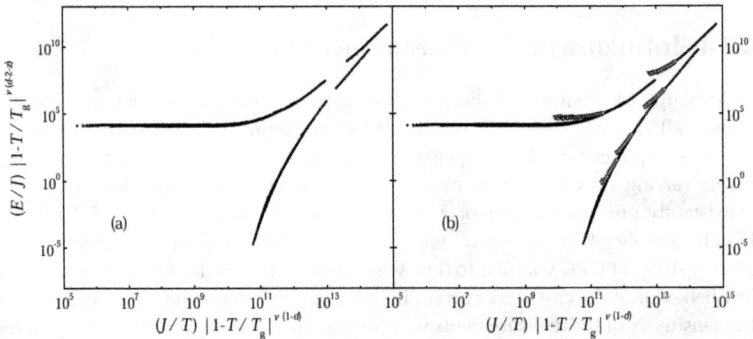

Figure 4.8: Glass scaling for the same data of Fig. 4.7 (Y121, $\mu_0 H$ = 1.0 T) but different scaling parameters: T_g' = 85.65, z' = 6.1, ν' = 1.5. (a) With the low electric-field range reduced to $E \geq$ 10^{-5} V/m (solid dots), one perceives apparently excellent scaling quality fully comparable to that of Fig. 4.7. (b) However, with the low field data (10$^{-5} \geq E$) included, for these parameters the scaling breaks down due to the visible deviations from the branches of the universal scaling function (open triangles). Therefore, even a seemingly good scaling in a reduced electric-field range cannot be deemed conclusive evidence for extracted dynamic exponents.

fields compared to many other experiments) and thus excluding the low field part (E < 10^{-5} V/m) of the I-V curves leads to a considerable broadening of the range of suitable scaling parameters: to higher glass temperatures and, especially, to lower dynamic exponents. Naturally, the original set of scaling parameters (T_g = 84.55 K, z = 9.1, v = 1.65) still yields excellent results but for the reduced window, corresponding to solid dots in Fig. 4.8(a), a scaling of comparable quality could also be achieved for T_g' = 85.65 K, z' = 6.1, and v' = 1.5. In fact, one can obtain scalings of such quality for sets with parameters ranging from the original to the primed values. This is not in contradiction to reports of the sensitivity of the glass scaling procedure: while Wöltgens et al. demonstrated that the scaling reacts very sensitively to the detuning of any *single* of the three parameters (with the other parameters fixed) [Wölt95], appropriately adjusting all three parameters simultaneously can obviously still produce a scaling of apparently equal quality, provided the measurement window is small enough. However, if the sensitivity is extended to lower electric fields the range of suitable parameters is narrowed. As illustrated by the open triangles in Fig. 4.8(b) indicating the available low field data (E < 10^{-5} V/m), the breakdown at low electric fields of the scaling with the low dynamic exponent, z' = 6.1, is apparent: for temperatures above and below T_g clear deviations from the universal branches appear due to the detectable crossover from asymptotic to linear (ohmic) behavior at lower current densities. The dependence of the range of suitable dynamic exponents on the measurement window should not be mistaken for an electric-field dependence of z.[21] The dynamic exponent of the glass transition in YBCO does not actually change with the electric-field window in this study; only the *range* of suitable values changes, unambiguously yielding the correct exponent only for large ranges in E extending to low electric fields.

4.4 Resistive Transitions

The analysis of resistive transitions $\rho(T)$ taken at various fixed magnetic fields presents a further possibility to corroborate the results obtained from the current voltage characteristics. Such transitions were recorded for the same magnetic fields $\mu_0 H$ = 0.03–1.0 T as the CVC's with a constant current density of 5×10^4 A/m² ($\cong$ 1 μA) well below J_0^+. The most appropriate approach for an analysis in the context of the vortex glass model is the Vogel-Fulcher relation. All data was plotted as $[d(\ln\rho)/dT]^{-1}$ vs T and a linear fit to the region of the critical transition $T_g < T < T_f$ was performed. With the procedure explained in Sec. 3.4 the glass temperature T_g, the product of the critical exponents $v(z-1)$, and the fluid temperature T_f could be determined. Within the experimental uncertainty all data obtained for the various magnetic fields, compiled in Tab. 4.1, fully match the results of the analyses of the CVC's.

---

[21] See, for example, [Lund98] where CVC's for a Thallium compound were analyzed at relatively high voltages— possibly in the FFF regime—and different values for T_g and z were found.

As explained in Sec. 3.4, this analysis contains potentially large uncertainties, especially for the critical exponents. Thus it should not be used as principal evidence for a vortex glass state, especially if one considers that, by definition, the system is investigated only at temperatures well above T_g. However, the good agreement for all results as determined from resistive transitions and CVC's—the temperatures T_g and T_f as well as the product $\nu(z-1)$—certainly gives further support to the validity and consistency of the presented analyses. Moreover, even though the critical exponents cannot be determined individually in order to confirm the increased dynamic exponent directly, the product $\nu(z-1) \approx 12\pm2$ obtained from $\rho(T,B)$ also strongly sustains a value of $z > 6$ rather than $4 \leq z \leq 6$. In the latter case, a static exponent of $2.4 \leq \nu \leq 4$ would result, in clear contradiction to the present data, previously published reports, and theory.

4.5 Alternative Models

While all of the above results present strong evidence for a second order vortex glass transition with a high dynamic exponent, there exist a variety of influences within the VG theory, mostly experimental artifacts such as sample geometry and morphology, as well as completely different models which could be construed to present an alternative explanation for the experimental data. However, in light of the different analyses—in particular that of the crossover current—the notion of experimental artifacts can be refuted, and alternative theories fail in consistently explaining the entire data.

Finite size effects have been reported to result in reduced glass temperatures and increased dynamic exponents $z > 6$ and could thus be surmised to constitute the origin of the present results as well. Since the (3D) vortex glass correlation length ξ_{VG} diverges as $(T - T_g)^{\nu(z-1)}$, close enough to T_g it must inevitably exceed the dimensions of any finite sample. Consequently, the crossover current density J_0^+ cannot decrease continuously but will be cut off at $J \approx kT/\Phi_0 d^2$ (2.23), where d corresponds to the limiting sample dimension. Since $T - T_g \ll T$, an approximately constant crossover current density will be observed for all isotherms affected by the finite size of the sample. In the CVC's this will have to appear as a visible change in the power-law behavior of (2.29). Also, the temperature dependence of J_0^+ (4.2) and the correlation length (2.21) (cf. Figs. 4.5 and 4.6) must reveal this cutoff. Finally, the vortex glass scaling of (4.3) and (4.4) will also be degraded [Wölt95]. Such a behavior with $z > 6$ due to finite size effects has been observed only in extremely narrow measurement bridges with strip widths $w < 2\,\mu m$ [Ando92] and in thin films with thicknesses $t \ll 400\,nm$ [Dekk92b, Wölt95, Sawa98]. In particular, $z > 8$ resulted only for $t < 20\,nm$, and an anisotropic 3D scaling model was applied to explain these results [Sawa98].[22] The dimensions of the samples at hand, $w = 50\ \mu$m and $t = 400$ nm, are clearly outside these reported limits. As the electric-field sensitivity of the present measurement is considerably increased compared with previous experiments one could suspect a heightened sensitivity to finite size effects and thus an increased sample size limit. However, the strict accordance of J_0^+ and ξ_{VG} to the VG theory throughout the entire measurement range and the excellent scaling achieved conclusively rejects finite size effects as the origin of the increased dynamic exponent.

Even though the x-ray analysis indicated excellent sample quality and the critical exponents of a vortex glass scaling should be independent of the material properties of a particular sample one could still suppose that inhomogeneities in the bridge might yet affect transport measurements, especially when conducted on a very long bridge extending over an area of more than 4×4 mm^2. Such effects should be visible in a resistive transition as a shoulder, residual resistance at low T, or at least a large transition width. Yet, the measurements show only a steep drop of the resistance down below the experimental resolution over more than six decades within 0.5 K. In addition, the analysis of the restricted high electric-field window ($E > 10^{-6}$ V/m) nicely reproduces the results of previous measurements on shorter bridges. Also, for microstructural properties found in YBCO which can effect the pinning in the sample only a decrease of z has been reported [Naki94]. Hence, there is no indication that inhomogeneities or inferior sample quality may be the cause of the high dynamical exponent z.

With respect to correlated (linelike) disorder Fisher and later Nelson and Vinokur have suggested the existence of a Bose glass (BG) phase in YBCO [Fish89b, Nels92, Nels93], which could present an explanation for deviating critical exponents. While for samples with induced columnar defects, e.g. created by heavy ion irradiation, a Bose glass has been observed [Wort88, Tink96], there have been contradicting reports as to its existence in pristine (unirradiated) YBCO samples [Silv93, Wölt93]. Besides, a BG scaling on the present data yields even larger deviations for the critical exponents, $z_{BG} = \frac{1}{2}(3z_{VG} + 1) \approx 14$ and $\nu_{BG} = \frac{2}{3}\nu_{VG} \approx 1$ (instead of $z_{VG} \approx 9$ and $\nu_{VG} \approx 1.5$), which does not render this interpretation plausible. However, in view of a recent report on the importance of screw dislocations (i.e. linelike defects) for pinning processes in YBCO thin films [Dam99] a further investigation of this aspect seems desirable. A definite exclusion of a possible Bose glass phase will require measurements with the magnetic field oriented at different angles from the c axis.

---

[22] In this model different correlation lengths and static exponents exist within the ab plane ($\xi_{\parallel}$ and $\nu_{\parallel}$) and along the c axis ($\xi_{\perp}$ and $\nu_{\perp}$). Thus, in equations (2.23a), (4.2), and (4.3) the terms ξ_{VG}^{1-z}, $\nu(z-1)$, and 2ν are replaced by $\xi_{\parallel}\xi_{\perp}^{-z}$, ($\nu_{\perp}z-\nu_{\parallel}$), and ($\nu_{\perp}+\nu_{\parallel}$), respectively. An interpretation of the results according to this anisotropic model fails as the static exponent $\nu \approx 1.2–2.0$ would increase to $\nu_{\perp} \approx 3–4$.

The model of flux creep also fails to account for the experimental observations. Although there may exist a cutoff of the vortex glass behavior *outside* (i.e. below) the measurement window, where other processes such as FC or TAFF may dominate, the entire data strictly adhere to the qualitative predictions of FFH. More specifically, it was recently suggested by Wen *et al.* that VG may exist only at high current densities, while for lower J the flux motion will change from collective creep or glassy motion to TAFF [Wen97]. This was concluded from a combined transport and magnetization measurement, where the transport CVC's (limited to $E \geq 10^{-3}$ V/m) yielded a glass temperature of 82 K and negative curvature for I–V curves below T_g, all of which exhibited positive curvature at lower J in the magnetization CVC's (10^{-7} V/m $\leq E \leq$ 10^{-5} V/m). The origin of the transition to TAFF for decreasing J is seen in the optimal hopping length of the flux-line activation process (cf. Chap. 3) exceeding sample dimensions, i.e. finite size effects. This assumption is actually supported by the behavior of the crossover current visible in their CVC's (cf. Fig. 1 of [Wen97]) which shows non-VG nature (J_0^+ increases with decreasing T). Also, within the context of the analysis presented in this chapter, their $T_g \approx 82$ K extracted from the high electric-field transport CVC's appears overestimated due to the onset of FFF at high E. Hence, their suggested transition is no intrinsic property of the superconducting system but a result of the specific sample geometry and can be explained within the VG model. In particular, while the FC model has been shown to result in a negative curvature in double logarithmic CVC's similar to that of a vortex glass phase, it fails to explain the strict power-law behavior of the crossover current density. Furthermore, it has been claimed that a vortex glass scaling should result in $z \sim 15$ and $\nu < 1$ if flux creep dominated vortex dynamics [Copp90, Wen97], which is in clear contradiction to the present experimental data and thus rejects the flux creep hypothesis.

4.6 Conclusions on the Vortex Glass Phase in YBCO

Based on the analyses of the current voltage characteristics and the resistive transitions the complete data can be consistently interpreted in terms of a second order transition from a glassy to a fluid state at the glass temperature T_g, followed by a change from an interacting fluid to free flux flow at the fluid temperature T_f throughout the magnetic-field range investigated. The resulting B-T phase diagram of Fig. 4.9 with the glass and fluid transitions given as $B_g(T)$ and $B_f(T)$, respectively, reveals the expected behavior for such a transition. In particular, it is possible to fit the temperature dependence of the glass line $B_g(T)$ according to the established relation $H_g(T) = H_0(1 - T_g/T_c)^{3/2}$, indicated by the solid line, where $\mu_0 H = 83.5$ T in very good agreement with previous results [Kötz94a, Kötz94b]. The transition is characterized by static and dynamic critical exponents ν and z, respectively, which exhibit a clear magnetic-field dependence for low B. Despite the onset of this dependence at a somewhat larger value of $\mu_0 H \sim 0.3$ T it qualitatively agrees with

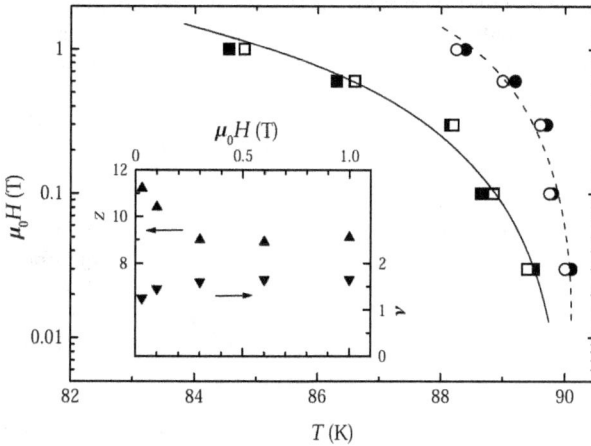

Figure 4.9: *H–T* phase diagram presenting the variation of the glass temperature T_g (squares) and the fluid temperature T_f (circles) with magnetic field as determined from the critical scaling (open symbols) and Vogel-Fulcher analyses (solid symbols). The solid line is a fit to $H_g(T) = H_0(1 - T_g/T_c)^{3/2}$; the dashed line is a guide to the eye. The inset reveals the magnetic-field dependence of z and ν below 0.3 T.

prior reports of increasing of z at low magnetic fields ($\mu_0 H \leq 0.05$ T) [Robe94, Noji96]. Possible interpretations for this behavior include the suppression of correlation effects—which the vortex glass theory is based upon as opposed to single vortex pinning models—due to the reduced flux-line density at these fields. As the increased intervortex distance may reduce the number of vortices within a glass correlation length, the underlying dynamics of the transition may change fundamentally. A conclusive explanation for this phenomenon has yet to be presented.

However, the most important aspect of the experimental results is, without doubt, the excellent agreement of all analyses establishing a dynamic exponent $z \geq 9$ in the entire magnetic-field range. Especially the extended analysis of the crossover current and the excellent results for vortex glass scaling of CVC's spanning more than 7 orders of magnitude in electric field down to below 10^{-8} V/m give very strong evidence for this conclusion. The consistent determination of the *I–V* glass line and the Vogel-Fulcher analysis of the resistive transitions further corroborate these results. Although the obtained value of z clearly exceeds values reported previously from transport measurements one should note that those results (typically 4–6) can be explained within the framework of the present analysis as a consequence of the restricted measurement window: the high dynamic exponent can be unambiguously identified only at very low electric fields.[23]

---

[23] The good agreement of T_g with theory and previous results is easily understood in terms of the comparatively small error in T_g ($\delta T_g < 5$ %) introduced by the high electric-field window in contrast to the error in z, where $\delta z \approx$ 50 %.

While the extracted $z = 9$ exceeds the quantitative theoretical predictions of FFH (4–7) all data qualitatively agrees with the vortex glass model and alternative explanations other than a VG transition with a high dynamic exponent fail to account for the observations. Specifically, the influence of finite size effects was unambiguously excluded based primarily on the consistent VG behavior of the crossover current but also on the good 'conventional' ($z \approx 6$) scaling of the data from a restricted electric-field window. Both, the models of Bose glass and of flux creep, are not able to account for the experimental observations and must be excluded as possible explanations for the increased dynamic exponent. Thus, while the characteristics of a vortex glass transition are clearly present, the deviation of z from theoretical predictions raises some issues, for instance the universality class of this glass. Originating from the XY spin glass, the VG model does not take into account various aspects particular to the vortex system, such as the normal conducting core. The dimensionality and specific dynamic behavior of the system, e.g. slow vortex line crossing and recombination [Fish89], although not changing the general character of the glass transition, may well have an influence on the value of parameters describing this transition, including critical exponents.

Clarification of these questions will require further theoretical and experimental efforts. For the latter the extension of the electric-field sensitivity to yet lower regimes appears promising as the presented technique of long measurement bridges still holds substantial potential in this direction: a bridge of length ~60 cm was successfully prepared in the context of this work but could not yet be experimentally investigated. Successful transport measurements conducted on this sample after the completion of this work documented a further increased electric-field sensitivity [Bass99a] thus opening up the perspective of bridge lengths of 1 m and more. Furthermore, a combination of CVC transport measurements with SQUID magnetization measurements, similar to those of [Wen97], could provide a possibility to extend the accessible electric-field range in *one* sample by several orders of magnitude and to check the consistency of the results in different measurement types. Regarding the dynamic aspects of the vortex glass transition, ac susceptibility measurement [cf. Kötz94b] may give further insight into the question of high dynamic exponents.

5 The Vortex Instability at High Dissipation Levels

Vortex motion at low velocities and low dissipation levels, as discussed in the previous chapters, has been at the focus of interest in the study of the vortex dynamics in high-T_c superconductors during recent years. In particular, the competition of the various mechanisms involved—thermal activation, pinning, inter-layer coupling, and collective effects due to vortex-vortex interactions—was studied in the range where the respective energies are of the same order and no single contribution dominates, such as the onset of dissipation at low driving forces. In the opposite limit of large Lorentz forces, where the vortex dynamics are governed by transport currents rather than pinning or thermal energies, the high resulting vortex velocities and increased dissipation present the opportunity to investigate a variety of different mechanisms and aspects of high-temperature superconductivity. In some high-T_c superconductors an interesting electronic instability has been observed and was explained in terms of a vortex instability predicted by Larkin and Ovchinnikov (LO) in 1975 [Lark75]. Their theory is based on free single vortices subjected to a driving (Lorentz) force and a damping force due to viscous drag resisting the vortex movement, as described in the theory of free flux flow [Tink64, Bard65]. The instability resulting at high current densities is related to the relaxation rate of quasi-particle energy. Inelastic quasi-particle scattering is not only relevant to the interpretation of experiments on thermal conductivity [Yu92], infrared optics [Nuss91], and high-frequency conductivity [Bonn93, Gao93] but also to the theoretical treatment of the electronic structure and mechanism behind high-temperature superconductivity [Pick92, Newn93]. Following the first experimental observations of the instability in the high-temperature superconductors YBCO [Doet94] and BSCCO (cf. Chap. 6 and [Xiao98a]) this topic has attracted increasing attention in the last years [Anto99, Decr99, Paul99]. The following chapter will briefly discuss the model of FFF and the vortex instability in the framework of the LO theory. It will conclude with the extension to this theory by Bezuglyj and Shklovskij (BS), which accounts for the influence of inevitable quasi-particle heating [Bezu92] and builds the basis for the subsequent analysis of the experiments.

71

5.1 Flux-Flow Dissipation

Soon after the theoretical prediction [Daun46, Ginz53, Bard56] and experimental verification [Cora54, Bion56, Glov57] of an energy gap Δ on the order of kT_c between the ground state and the quasi-particle excitations of a superconductor, the pairing theory of Bardeen, Cooper, and Schrieffer (BCS) was presented [Bard57], giving a microscopic theory of superconductivity. The essential idea behind the BCS theory is the creation of bound pairs of electrons, created from the Fermi-sea ground state of the electron gas due to a weak attraction between electrons arising from the second-order electron-phonon interaction. The minimum energy needed to break up one of these "Cooper pairs" (producing two quasi-particle excitations) increases from zero at T_c to a limiting value at $T = 0$ given by $2\Delta(0) = 3.528kT_c$. Such a creation of two quasi-particle excitations and the reverse process of a recombination to a Cooper pair is dominated by electron-phonon scattering, whereas the direct interaction between quasi-particles relies on electron-electron scattering. The density of Cooper pairs is essentially given by the superconducting order parameter (analogous to the GL relation $n_s \propto |\psi|^2$) and low energy bound states of the quasi-particle can exist inside a vortex, where the energy gap is reduced. A detailed introduction to the BCS theory can be found in [Genn66, Tink96].

For the instability phenomena to be investigated in the following chapters the process of power dissipation in a superconductor as treated within the framework of the BCS theory is most interesting. Whenever the Lorentz force due to a transport current exceeds the effective pinning force, flux lines will essentially be able to move freely through the superconductor, which has lead to the term free flux flow. The force resisting a continued acceleration of the vortices will emanate from viscous drag, comparable in principle to the velocity dependent damping force acting on moving solid objects in viscous or gaseous media. Two mechanism have been proposed in this context. First, as mentioned in Sec. 2.3.1, a moving vortex, representing a varying magnetic field, will induce an electric field opposing the external transport current and give rise to a dissipative voltage. An additional mechanism of dissipation relies on the changed density of Cooper pairs inside the core of a vortex. In front of a moving vortex the Cooper pair density will have to be reduced from the maximum value outside (where $\psi \approx \psi_{max}$) according to the depression of the order parameter inside the vortex core. Cooper pairs must thus break up in front of the vortex and recombine behind it. Depending on the vortex velocity, i.e. the necessary rate of quasi-particle excitation and recombination, the relaxation time τ characterizing these processes will play a crucial role. Only if the vortex velocity is low enough to allow the change in Cooper pair density at the front and back of the flux line via equilibrium states, the energies required for excitations and released in recombinations will be equal.

In 1964 Tinkham introduced his model for dissipation in the flux-flow regime. Based on the absence of pinning (equivalent to negligible pinning energies at suffi-ciently high driving forces), the model expects damping of vortex motion by viscous drag only. Specifically, assuming a viscous drag coefficient η such that the resulting force per unit length of a vortex line moving at a velocity v_L is

$$f_\eta = -\eta v_L \qquad (5.1)$$

and equating this expression to the Lorentz force (2.9) $f_L = J\Phi_0$ one obtains the flux-flow resistivity

$$\rho_f = B\Phi_0/\eta \qquad (5.2)$$

corresponding to the first mechanism mentioned above. The obtained expression for the resistivity is in good agreement with the empirical relation

$$\rho_f/\rho_n \cong H/H_{c2} \qquad (5.3)$$

for high magnetic fields $H \gg H_{c1}$ only [Hemp64]. For low magnetic fields on the other hand, the mechanism of quasi-particle excitation and recombination becomes more important.

Tinkham first suggested a finite relaxation time τ such that $\tau^{-1} \approx \Delta(0)/\hbar$ at $T = 0$ and $\tau^{-1} \propto (1 - T/T_c)$ near T_c, based on the conjecture that the relevant scale will be given by the time needed for a quasi-particle to diffuse over a length $\sim\xi$ [Tink64]. Bardeen and Stephen later presented a model of a local superconductor with a fully normal core of radius $\sim\xi$ [Bard65]. Although $\psi(r)$ strictly vanishes only for $r = 0$ and rises roughly as $\psi_\infty r/\xi$ within the core region, in their calculation of a quasi-particle spectrum Caroli et al. were able to show that such a structure yields a density of low-lying excitations in the vortex [Caro64, Genn66] which is essentially the same as that of a normal cylinder of radius $\sim\xi$ [Tink96]. For high magnetic fields $H \lesssim H_{c2}$, where the distance between the vortices is of the same order as their diameter, this model will fail as the quasi-particles do not remain localized to a specific vortex—similar to conduction electrons in the periodic potential of an atomic lattice. For magnetic fields not too close to H_{c2}, however, the model is justified and the total dissipation in the core is obtained

$$W_d = v_L^2 \Phi_0^2 / 2\pi\xi^2 \rho_n \qquad (5.4)$$

Comparing this result with the dissipation per unit length of flux line due to viscous damping $W = -f_\eta \cdot \bar{v}_L = \eta v_L^2$ one can obtain an expression for the damping coefficient $\eta \approx \Phi_0 H_{c2}/\rho_n$. The latter, when combined with (5.2), leads directly to

$$\rho_f/\rho_n \approx B/H_{c2} \qquad (5.5)$$

5.2 Larkin-Ovchinnikov Instability

Whereas the Bardeen-Stephen model assumes that quasi-particle excitations are not able to leave the region of the vortex core during their lifetime, $v\tau \ll \xi$, Larkin and Ovchinnikov (LO) have explored the dissipative processes in the opposite limit of high flux-flow velocities [Lark75]. In particular, they predicted an instability of the vortex system at a critical vortex velocity v^*, which should manifest itself as a voltage jump at high current densities in I-V curves of type II superconductors. The key aspect behind the mechanism leading to the instability is the particular velocity dependence of the damping coefficient $\eta(v)$.

The linear dependence of the damping force $f_\eta \propto v$ in the Bardeen-Stephen model implies a monotonous increase of the damping force with vortex velocity. However, at high velocities and dissipation levels the assumptions leading to (5.3) may not be valid anymore, especially since the motion of the particles will not be restricted to the core of the vortex. In fact, Larkin and Ovchinnikov showed that for sufficiently large τ and v the rate of excitation will exceed that of recombination and quasi-particles can effectively escape from a moving vortex. In their calculation of the change of the quasi-particle energy distribution due to an excitation energy ϵ, which depends on the electric field accelerating the particles, they found a decrease of the density inside the vortex only for $\epsilon \gg \Delta(0)$. Thus, for high electric fields (i.e. large vortex velocities v) the loss of quasi-particles changes the spatial variation of the order parameter [Gork73]

$$\Delta^2(r) = \Delta_\infty^2 r^2 / \left[r^2 + \xi^2(v) \right] \tag{5.6}$$

The velocity dependence of the coherence length is given by

$$\xi^2(v) = \xi^2(0) / \left[1 + (v/v^*)^2 \right] \tag{5.7}$$

where $\xi^2(0) = \pi D / 8(T_c - T)$, D is the diffusion coefficient, and v^* a critical velocity [Lark75]. Hence for larger v the coherence length and thus the core diameter will shrink. According to [Lark71, Gork73] the damping coefficient $\eta(v)$ can be obtained from the ratio of the core areas of the moving and the stationary vortex, corresponding to the respective number of quasi-particles, as

$$\eta(v) = \eta(0)\xi^2(v)/\xi^2(0) \tag{5.8}$$

with a temperature dependent value of $\eta(0) \approx 0.45\sigma_n T_c / D(1 - T/T_c)^{1/2}$ where σ_n is the normal state conductivity. Combining (5.7) and (5.8) and plotting η and f_η vs the normalized vortex velocity (Fig. 5.1) reveals a damping coefficient that decreases considerably above $v \sim 0.2\, v^*$ and approaches zero for high vortex velocities $v \gg v^*$. The damping force displays an almost linear increase in the low velocity regime but a

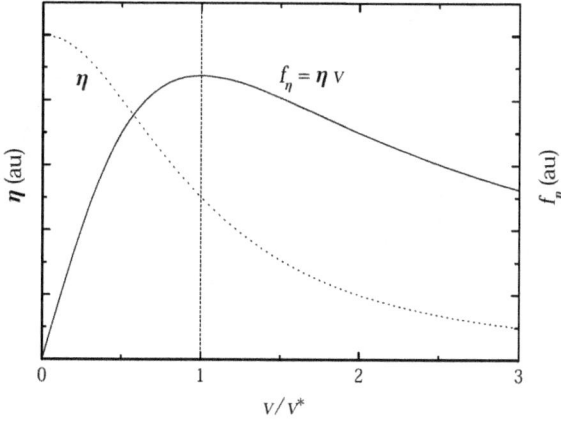

Figure 5.1: Velocity dependence of the damping coefficient η (dotted line) and the damping force f_η (solid line) according to the LO theory. The non-monotonous behavior of $f_\eta(v)$ with a maximum at the critical vortex velocity v^* is essential for the appearance of the vortex instability.

maximum at $v = v^*$ and a subsequent decrease at higher velocities. Within the LO theory the critical velocity is given by

$$v_{LO}^* = 1.02(D/\tau_{LO})^{1/2}(1 - T/T_c)^{1/4} \qquad (5.9)$$

with the corresponding electric field induced by a moving vortex ensemble $E_{LO}^* = v_{LO}^* B$.

Obviously, if a transport current is increased beyond a critical current density J^* and the vortices reach a velocity exceeding v^* the decreasing damping force will no longer be able to compensate the driving Lorentz force and the flux lines will begin to accelerate, resulting in a further decrease of the damping force, which in turn will increase the acceleration, which again decreases the damping force, and so on and so forth. The sudden increase of vortex motion appears as a voltage jump which indicates the instability of the vortex ensemble at v^* in I-V curves, illustrated schematically in Fig. 5.2. Extracting the parameters characterizing this instability, in particular the value of the critical electric field E_{LO}^*, allows to determine the quasi-particle scattering rate $1/\tau$ if the diffusion coefficient can be obtained from independent experiments.[24]

---

[24] The LO theory gives no explicit predictions regarding the behavior of η at very high vortex velocities $v \gg v^*$. While the obtained $f_\eta(v)$ suggests vanishing damping forces for extremely high v and thus an acceleration of the vortices until the dissipation in the sample leads to a loss of superconductivity, there may be a different 'secondary' mechanism impeding vortex motion for $v \gg v^*$.

75

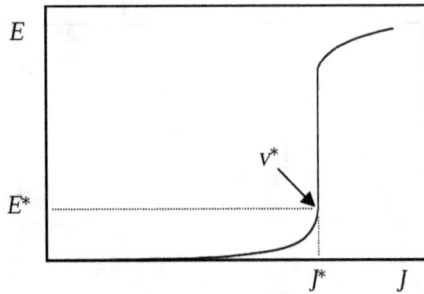

Figure 5.2: Schematic representation of an I-V curve with the vortex instability appearing as a jump in the electric field at J^* where the vortex velocity reaches the critical value v^*.

5.3 Bezuglyj-Shklovskij Extension for Quasi-Particle Heating

A fundamental assumption of the LO model, which neglects electron-electron interaction in the quasi-particle processes, is the identification of the total inelastic scattering rate $1/\tau_{in}$ with that of electron-phonon interaction $1/\tau_{ep}$, which essentially implies that the excitation energy of the quasi-particles is removed directly to the crystal lattice of the superconductor instead of thermalizing the quasi-particle distribution. However, Bezuglyj and Shklovskij (BS) were able to show in their theoretical work that the finite rate of removing the dissipated power in the sample will notably affect the instability [Bezu92].

The heat removal process occurs in two stages: from the quasi-particles energy is transferred by radiation of non-equilibrium phonons; from the crystal lattice of the sample heat flows to the substrate by phonon exchange through the interface. Considering in the latter process the free path length of phonons with respect to scattering by electrons $l_{ep}(T) \approx \hbar v_F / kT$ (with v_F the Fermi velocity) and the so-called effective film thickness $d_{ef} = d/\alpha$ (with the film thickness d and the mean probability $0 < \alpha < 1$ of phonon transmission to the substrate) BS distinguish two cases: $l_{ep} \ll d_{ef}$, where emitted phonons are reabsorbed and both systems (quasi-particles and crystal lattice) have a higher temperature than the substrate (Joule heating regime); and the opposite limit of $l_{ep} \gg d_{ef}$, where non-equilibrium phonons leave the sample and the heat removal rate is determined by the value of the electron-phonon coupling constant (electron overheating regime). Nonetheless, in both cases the heat removal rate remains finite. As the LO vortex instability depends fundamentally on the non-equilibrium energy distribution it is thus reasonable to expect this inevitable heating of quasi-particles to influence the particular appearance of the instability.

BS investigated the consequences of this effect based on the heat flow from quasi-particles to bath

$$T_{qp} - T_{bath} = d\sigma E^2/h \qquad (5.10)$$

with the (total) heat-transfer coefficient h and the sample conductivity σ. Together with the CVC extremum condition

$$\frac{d}{dE}[\sigma(E)E]_{E=E^*} = 0 \qquad (5.11)$$

they were able to obtain analytical parametric solutions for the threshold electric field E^* of the instability for all temperatures and magnetic fields

$$\frac{E^*}{E_0} = (1-t) \cdot \frac{3t+1}{2\sqrt{2}t^{3/4}(3t-1)^{1/2}} \qquad (5.12)$$

The parameter t depends on the applied magnetic field B and a characteristic magnetic field B_T. It is given by

$$t = \frac{T_c - T_{qp}}{T_c - T_{bath}} = \frac{1+b+(b^2+8b+4)^{1/2}}{3(1+2b)} \qquad b = \frac{B}{B_T} \qquad (5.13)$$

The normalization parameter E_0 depends on B_T, the quasi-particle diffusion coefficient D, and the inelastic relaxation time τ_{in}

$$E_0 = 1.02 B_T (D\tau_{in})^{1/2}(1 - T/T_c)^{1/4} \qquad (5.14)$$

Similarly, an expression for the critical current density $J^* = \sigma(E^*)E^*$ was obtained

$$\frac{J^*}{J_0} = \frac{2\sqrt{2}t^{3/4}(3t-1)^{1/2}}{3t+1} \qquad (5.15)$$

with an analogous normalization current density

$$J_0 = 2.62(\sigma_n/e_0)(D\tau_{in})^{-1/2} kT_c (1 - T/T_c)^{3/4} \qquad (5.16)$$

One notes that $t = 1$ (i.e. $B = 0$) yields $E^* = 0$ and $J^* = J_0$, which represents the limiting critical current density for vanishing magnetic fields. The physical meaning of the characteristic magnetic field

$$B_T = 0.37 \frac{e_0 h}{k} \cdot \frac{\tau_{in}}{\sigma_n d} \qquad (5.17)$$

becomes apparent from (5.13). In magnetic fields $B \ll B_T$, with a low vortex density and small heating effects, T_{qp} increases proportionally to B. In this case the difference $T_{qp} - T_{bath}$ is minimal and the vortex instability is dominated by the LO mechanism due to the non-equilibrium distribution of the quasi-particles. However, in the opposite limit of high magnetic fields $B \gg B_T$ the temperature T_{qp} approaches the constant value $T_c - \frac{1}{3}(T_c - T_{bath}) \geq \frac{2}{3} T_c$. Due to the weak heat removal the distribution of quasi-particles becomes thermalized and represents an equilibrium distribution of a higher temperature. Thus, the instability is mainly due to the influence of heating rather than the LO mechanism based on an actual non-equilibrium distribution. The latter manifests itself as a reduction in the critical electric field compared to the LO theory $E^* \ll E_{LO}^*$ for high magnetic fields.

6 Flux-Flow Instability in $Bi_2Sr_2CaCu_2O_{8+\delta}$

A large number of interesting physical phenomena is associated with super-conducting systems at high dissipation levels, including for instance the flux-flow instability [Lark75], recrystallization of the vortex ensemble [Kosh94], and self-organized criticality (SOC) [Pla91]. As they all depend on high flux-flow velocities of vortices, the typical experimental technique used to investigate these phenomena is the analysis of current voltage characteristics taken at high current densities and electric fields. In particular, the observation of voltage instabilities in CVC's has been used extensively in such studies as the easily observable jumps allow the rather precise determination of the corresponding parameters (compared with gradual changes in the characteristics, for example, associated with the crossover current in the vortex glass transition). Such discontinuities have been reported for a wide variety of systems, including the LTSC's Sn [Brem59, Musi80, Klei85], Al [Musi80, Klei85], In [Klei85], Mo_3Si [Samo95], Ta/Ge [Ruck97], Nb [Ando93], $NbSe_2$ [Hend96], as well as the HTSC system $YBa_2Cu_3O_7$ [Doet94, Xiao96]. Due to its low irreversi-bility field and the resulting extended flux-flow regime, $Bi_2Sr_2CaCu_2O_{8+\delta}$ could be expected to represent an excellent candidate for the observation of the flux-flow instability in large temperature and magnetic-field ranges. Furthermore, considering this system's strongly layered nature one should expect the investigation of the effect of anisotropy on this phenomenon to be particularly interesting. The following chapter will discuss the discovery of the vortex instability in the $Bi_2Sr_2CaCu_2O_{8+\delta}$ compound.

6.1 Sample Information and Experimental Details

Four BSCCO samples were used in the investigation of the voltage instability, two of which were studied systematically. Section 6.2, presenting measurements with the magnetic field oriented parallel to the c axis ($H \parallel c$), concentrates on sample P92 (++), which was part of the study on the influence of oxygen stoichiometry on the transport properties of BSCCO. The preparation process and sample characteristics were described in detail in Sec. 3.1. This sample, a c-axis oriented film of 400 nm thickness sputtered onto [100] oriented $SrTiO_3$, possessed a superconducting critical temperature (inflection point) $T_{c,ip}$ = 86.0 K and a normal state resistivity ρ_n = 66 $\mu\Omega$cm at 100 K. Silver contact pads were evaporated and a measurement bridge (100 × 10 μm^2) was prepared by photolithography and wet chemical etching, allowing the measurement of I–V curves in the standard four-point method. The second sample, T11, with a thickness of 320 nm and a normal state resistivity of ρ_n = 506 $\mu\Omega$cm at 100 K, used in the study of the angle dependent properties ($H \nparallel c$) of the instability as presented in Sec. 6.3, was prepared analogously [Wagn93] but without further annealing treatment [as had been necessary for the increase of oxygen content in P92 (++)]. Both samples revealed excellent crystallographic c-axis orientation in x-ray measurements and the resistive transitions betrayed no indication of inhomogeneities or other signs of inferior sample quality.

All transport measurements were conducted in a variable-temperature [4]He-cryostat with a temperature stability better than 0.05 K. Magnetic fields up to 6 Tesla could be applied and the sample holder rotated to continuously tilt the c axis of the sample to an angle $0° \leq \theta \leq 180°$ with respect to the magnetic field, while the direction of the current was always kept perpendicular to the magnetic field. All measurements were performed with rectangular current pulses of 1 s length and an interval time of 3 s between pulses. To avoid its destruction, the maximum voltage across the stripline was restricted to 1 V.[25] The transport current was supplied by a current source with a maximum output of 0.1 A, and a nanovolt meter was used to record the voltage across the sample.

6.2 The Voltage Instability in CVC's

Figure 6.1 displays high current density CVC's of sample P92 at a fixed temperature T = 77 K and various magnetic fields, 0.25 T $\leq B \leq$ 6 T. Within the experimental resolution, a discontinuous voltage jump at a well-defined critical current I^* (obtained from the last stable measurement point below the jump) is visible for $\mu_0 H \leq$ 1.25 T whereas the I–V curves become more smeared at higher fields and the jump disappears at fields above 3 T. For a fixed magnetic field, the I–V isotherms exhibit an analogous behavior with the jump being more pronounced at lower and suppressed at higher temperatures. Similar observations have been reported for conventional superconductors [Musi80, Klein85, Ando93] and YBCO [Xiao96]. Typical values of the instability current exceed 10 mA corresponding to a minimum current density of

---

[25] Nevertheless, several samples were inadvertently destroyed during the experiments due to a low breakdown electric field of BSCCO. Thus, a comparative study of the influence of oxygen stoichiometry on the vortex instability in the samples of the oxygen doping series of Chap. 3 was not possible.

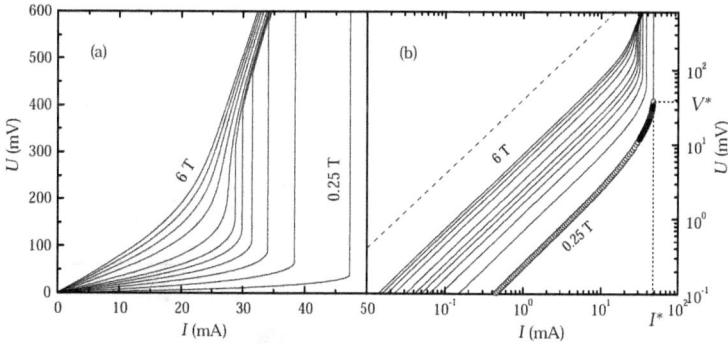

Figure 6.1: (a) Linear plot of the high dissipation CVC's of the $Bi_2Sr_2CaCu_2O_{8+\delta}$ sample P92 at $T = 77$ K and magnetic fields (from lower right to upper left) of 0.25, 0.50, 0.75, 1, 1.5, 2, 3, 4, 5, and 6 T. Definitions of the critical current I^* and the critical voltage V^* are indicated by the dotted line for one of the I–V curves. (b) Double-logarithmic plot of the same data. The flux-flow region with $V \propto I$ (indicated by the dashed line) is visible as an extended linear segment for all magnetic fields.

2.5×10^5 A/cm² for which an instability can be observed. For decreasing temperatures as well as for decreasing magnetic fields, V^* (the voltage below the jump) shifts to lower values while there is an increase in I^* (and in the voltage level of the I–V curve above the jump). The specific importance of this fact in relation to the LO theory will be discussed below; in regard to the measurements this primarily limits the range of accessible magnetic fields and temperatures for which data can be obtained as high voltages or current densities may destroy the sample due to excessive joule heating. One notes the extended region of FFF appearing as a straight I–V segment where $V \propto I$.

6.2.1 Heating Effects

Besides the vortex instability, a variety of different mechanisms have been associated with voltage jumps in I–V characteristics, including channel depinning [Hend96], lattice recrystallization, and SOC as well as thermal runaway and hotspots [Skoc74, Gure87, Xiao98b]. The influence of joule heating may not be neglected in the analysis of these measurements as the power dissipated in the sample below the instability, $P^* = I^*V^*$, is on the order of 1 mW. Depending on the value of P^*, on the effective heat conductivity of the sample in the experimental setup, and on the quality of the current contacts, superconductivity could be destroyed simply by excessive joule heating resulting in an abrupt temperature increase to above T_c and appearing as a sudden voltage jump in CVC's.[26] However, for a fixed temperature

---

[26] In this context, the possible limitation of heat conductance due to the substrate material has been noted [Doet95]. However, in the investigation of the instability in YBCO (Chap. 7) it will be shown that the

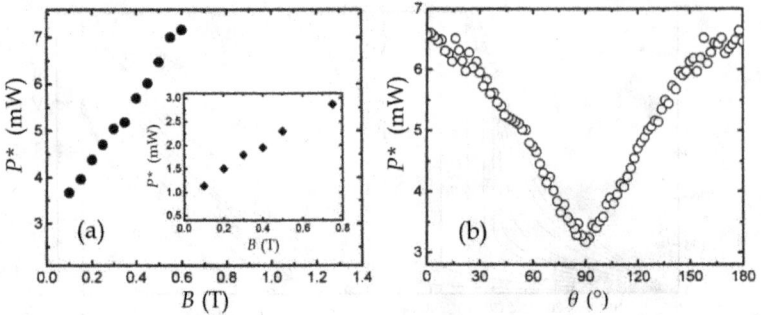

Figure 6.2: (a) P^* vs $\mu_0 H$ at various magnetic fields (for $\theta = 0°$) and fixed temperature for samples T11 ($T = 76$ K) and P92 (inset, $T = 79$ K) (b) P^* vs θ for sample T11 at $\mu_0 H = 0.5$ T and $T = 76$ K. Thermal runaway as a cause for the voltage jump cannot explain the dependence of the dissipated power before the jump on magnitude and orientation of the magnetic field.

the dissipated power should be independent of magnetic field if thermal heating is responsible for the voltage jump, which is clearly not the case in these measurements, as shown in Fig. 6.2. As $\mu_0 H$ is increased from 0.1 to 0.6 T the dissipated power doubled for both samples; angle-dependent measurements also revealed a strong dependence of P^* on the orientation of the sample in a constant magnetic field, with $P^*_{B\|c} \approx 2P^*_{B\perp c}$. Furthermore, experiments based on the experimental technique developed in Chap. 8 (conducted after the completion of this work) have shown only a minimal temperature increase in the microbridge due to dissipated power ($\sim$1 K) [Bass99b]. Thus, while the sample temperature will increase above the bath temperature, a stable thermal equilibrium is reached everywhere below the jump and thermal runaway can be excluded as the origin of the voltage instability.

Despite the minor increase in the average temperature of the stripline the dissipated power cannot be removed instantaneously and the creation of localized normal-conducting hotspots maintained by Joule heating may be possible. The hotspot effect, proposed as the explanation of the voltage jump phenomenon in some low-temperature superconductors, is characterized by a particular temperature dependence of the critical current $I^* \propto (1 - T/T_c)^\gamma$, with $\gamma = 1/2$ theoretically predicted and experimentally observed [Skoc74, Gure87].[27] Though, as it will be shown below, the critical current in these BSCCO samples clearly yields a different exponent $\gamma \approx 3/2$, which excludes both the hotspot effect and the crystallization of the flux-line system [Kosh94], where I^* increases with T (cf. Sec. 6.2.4). Also, the shape of the sharp single jump visible in the CVC's of Fig. 6.1 disagrees with the mechanisms of channel depinning [Hend96] and hotspots [Skoc74, Gure87, Xiao98b], which have

characteristic features of the instability remain unchanged both on $SrTiO_3$ and MgO substrates, in spite of the considerably different heat conductivities of these two materials. Although no conclusive proof by itself, this already supports the exclusion of thermal heating effects as the cause for the voltage instability in BSCCO as well.
[27] This effect has also been observed—with the same result, $\gamma = 1/2$—in samples of inferior quality of the high-temperature superconductor YBCO [Xiao98b].

been shown to result in stepwise instabilities. Finally, depinning and self-organized criticality may be rejected on the basis of the FFF behavior preceding the voltage instability [see Fig. 6.1(b)], as both of these mechanisms should occur at or near the vortex depinning current, i.e. below the flux-flow regime.

6.2.2 Larkin-Ovchinnikov Flux-Flow Instability

The LO theory is based on the assumption of Stephen-Bardeen free flux flow behavior. As mentioned above, the linear segments (with slope $s = 1$) of the I–V curves in Fig. 6.1(b) are characteristic for the ohmic temperature dependence of the resistivity in the FFF region. The linear magnetic-field dependence of the resistivity in the low current limit $\rho_f \propto B/B_{c2}$, presented in Fig. 6.3, confirms this FFF behavior. According to the LO theory the viscous force in this regime of vortex dynamics is a non-monotonous function of the flux-line velocity v. If the latter exceeds the critical value v_{LO}^*, for which the viscosity assumes its maximum, the retarding force on the vortices will begin to decrease resulting in the instability discussed in Sec. 5.2. As the voltage V across a sample of length l is related to the velocity v of the flux lines by

$$V = E \cdot l = Bvl \tag{6.1}$$

the critical velocity v^* can be determined from the voltage V^* directly before the sudden voltage jump in the CVC's of Fig. 6.1. Following (5.9) it is then possible to derive the inelastic scattering time of quasi-particles τ_{in} from v_{LO}^*. However, the extracted values of the critical current I^* (or current density J^*) and the critical

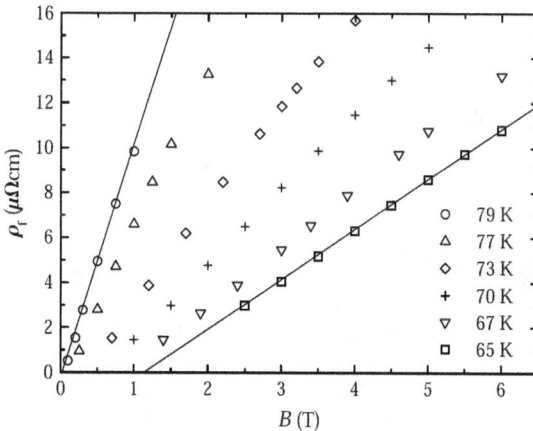

Figure 6.3: Linear magnetic-field dependence of the free flux flow resistivity ρ_f in the low current limit. Solid lines are guides to the eye.

velocity v^*, presented in Fig. 6.4, display a clear increase for low magnetic fields. This dependence on B has been observed experimentally but could not be explained within the LO model. By considering the finite removal rate of the power dissipated in the sample Bezuglyj and Shklovskij have recently shown in a theoretical work that this behavior can be explained in terms of the unavoidable heating of quasi-particles.

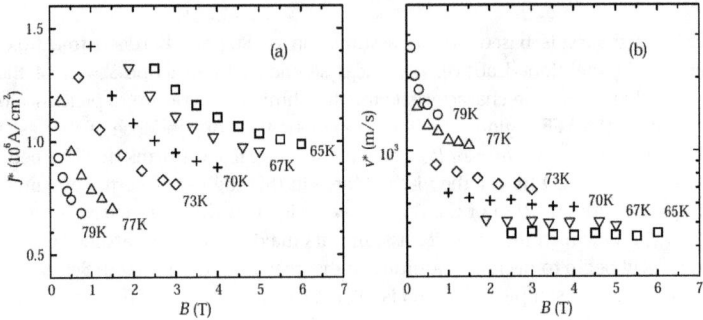

Figure 6.4: Magnetic-field dependence of (a) the critical current density and (b) the critical velocity at temperatures as assigned to each curve. Note the logarithmic scale for v^*.

6.2.3 Experimental Observation of Quasi-Particle Heating

Following the approach laid forth in Sec. 5.3, all data points for J^* and E^* were analyzed according to the parametric forms (5.12) and (5.15), with J_0, E_0, and B_T as free parameters. Plotting the normalized values of the critical electric field vs the critical current density for a given temperature

$$(E^*/E_0)=(1-t)/(J^*/J_0) \tag{6.2}$$

as shown in Fig. 6.5, will collapse the data of all magnetic fields onto a universal curve for suitably chosen J_0 and E_0. The symbols present the data for all temperatures and magnetic fields from Fig. 6.4; the solid line indicates the calculated universal curve (6.2). Rather good agreement between theory and experimental data is found, which corroborates the interpretation of this phenomenon as the Larkin-Ovchinnikov flux-flow instability. As illustrated in the inset of Fig. 6.5, the characteristic field B_T was extracted by plotting

$$(J^* E^*/J_0 E_0)=(1-t) \tag{6.3}$$

vs B/B_T, where the values of J_0 and E_0 from above were used and B_T was the only free parameter. Resulting values for B_T, ranging from 5 T at 65 K to 0.4 T at 79 K, are given in Fig. 6.6 along with those of J_0 and E_0, which also exhibit a strong temperature dependence. An analogous analysis of sample T11 yielded comparable results, with slightly increased values for J_0 and E_0 (arising presumably from the higher normal state resistivity and, possibly, also from the different oxygen content) and a somewhat higher B_T.

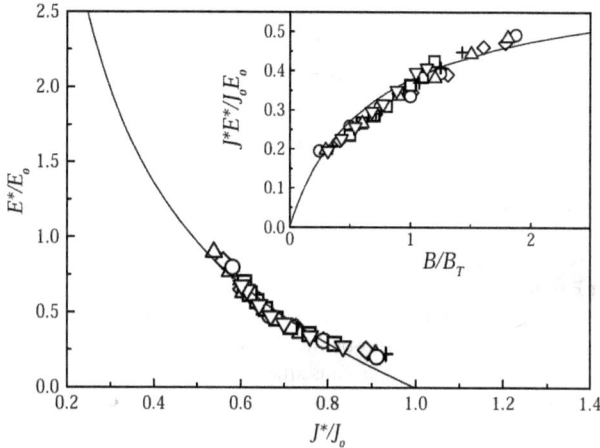

Figure 6.5: Scaling of the instability parameters J^* and E^* for all temperatures and magnetic fields to the universal curves of the BS model.

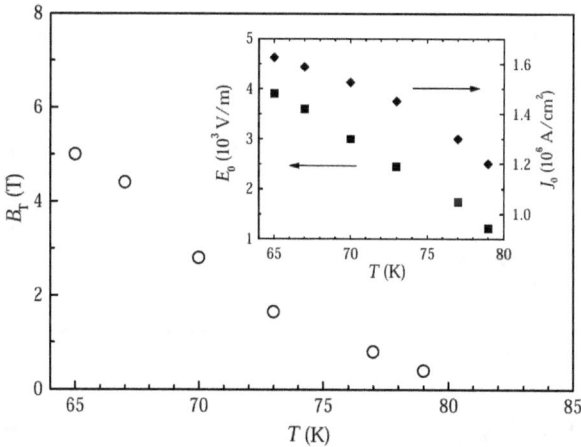

Figure 6.6: Scaling parameters E_0 and J_0 and the characteristic magnetic field B_T extracted from the BS analysis.

85

6.2.4 Quasi-Particle Scattering Rates and Diffusion Lengths

From the values of the fitting parameters the time τ_{in} of inelastic quasi-particle scattering, more commonly given as the quasi-particle scattering rate $1/\tau_{in}$, can be obtained along with the diffusion length l_e [Bezu92]. Extracted values for sample P92, given in Fig. 6.7, are somewhat lower than those reported for similar experiments on YBCO thin films [Doet94, Xiao96]; the results for T11 agree very well with published results. However, for both samples the qualitative temperature dependence of the scattering rate follows the same relationship. In the regime of inelastic electron-electron scattering, $1/\tau_{in}$ will be related to the quasi-particle density n_s and thus to the superconducting gap parameter $\Delta = 1.76kT_c(1 - T/T_c)^{1/2}$ [Doet94]. With $(1/\tau_{in}) \propto n_s^2$ and $n_s \propto \exp(-2\Delta/kT)$, one expects

$$1/\tau_{in} = 1/\tau_0 \exp(-4\Delta/kT) \tag{6.4}$$

where $1/\tau_0$ is the (extrapolated) value of the scattering rate at the critical transition temperature T_c. As shown in Fig. 6.8, the experimental data in the high-temperature range can be fitted with (6.4) yielding values for $1/\tau_0$ of 1.22×10^{11} s^{-1} and 1.71×10^{12} s^{-1} for P92 and T11, respectively. The slight increase of $1/\tau_{in}$ above the solid line for both samples at low temperatures can be explained in terms of the contribution of temperature independent electron-phonon scattering (cf. Sec. 7.2). The decreased scattering rate for P92 may be a consequence of the strongly increased oxygen content in this sample, although one would rather expect the higher charge-carrier

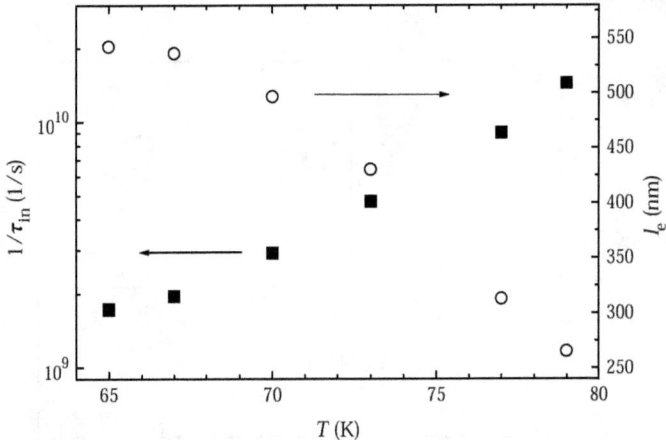

Figure 6.7: Temperature dependence of the inelastic quasi-particle scattering rate $1/\tau_{in}$ (solid squares) and diffusion length l_e (open circles) in sample P92.

density in P92 to result in a higher $1/\tau_0$ as well, if electron-electron scattering dominates. More theoretical and experimental work, such as a study of HTSC's with systematically varied oxygen stoichiometry, will be desirable.

Considering central assumptions of the LO theory—the quasi-particle excitations being able to escape from the vortex core and a spatial homogeneity of these excitations within the superconducting material—one finds the diffusion length l_e to be of special importance. Obviously, it must exceed the values of the vortex core dimension ξ and be at least on the order of the intervortex spacing a_0. As $\xi \lesssim 10$ nm and $a_0 \sim 50$ nm at $\mu_0 H = 1$ T the conditions are fulfilled for both samples, with $l_e \sim 400$ nm (P92) and ~ 150 nm (T11), given in Fig. 6.7 and in the inset of Fig. 6.8.

Also, both in low-T_c [Musi80, Klei85, Ando93, Ruck97] and high-T_c superconductors [Xiao96] a scaling behavior of the temperature and magnetic-field dependence of the critical current density J^* was found

$$J^*(T,H) = J^*(H) \cdot (1 - T/T_{c0})^{3/2} = J_0^* \frac{(1 - T/T_{c0})^{3/2}}{(1 + H/H_0)^\alpha} \tag{6.5}$$

where T_{c0}, H_0, and α are fitting parameters and J_0^* is the value of the critical current density at $T = 0$, $H = 0$. Both the temperature and magnetic-field dependence of J in the present samples [evident in Fig. 6.4(a)] scale with (6.5) and are shown in Fig. 6.9. For sample P92 one obtains $T_{c0} = 92$ K, $\mu_0 H_0 = 0.1$ T, $\alpha = 0.37$, and $J_0^* = 2.8 \times$

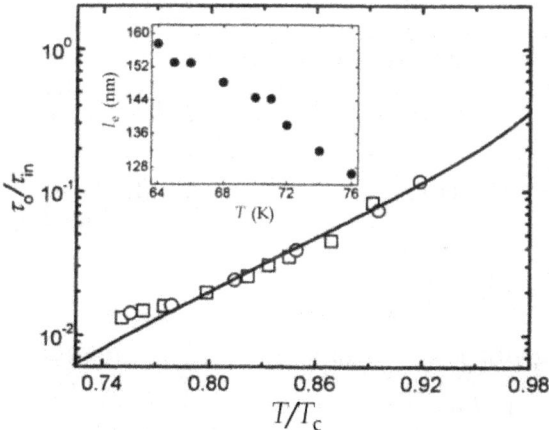

Figure 6.8: Fit of the scattering rates for samples P92 (open circles) and T11 (open squares) to the temperature dependence expected for electron-electron scattering. The data for both samples are well described by (6.4) indicated by the solid line. The inset presents the values of the diffusion length for sample T11.

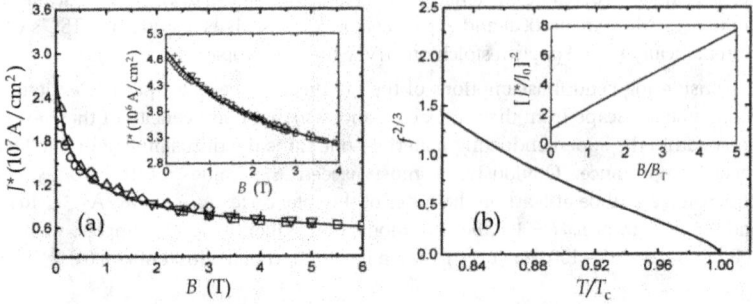

Figure 6.9: (a) Critical current densities for both samples scaled according to (6.5). (b) Approximated power-law behavior of (6.6) extending over the high-temperature range below to T_c.

10^7 A/cm²; for T11 $T_{c0} = 98$ K, $\mu_0 H_0 = 1.35$ T, $\alpha = 0.33$ and $J_0^* = 5 \times 10^6$ A/cm². Solid lines in Fig. 6.9 represent $J^*(T,H)$ calculated with (6.5) and the above values. Within the framework of the BS extension to the LO theory this scaling behavior results directly from the equations governing the critical current density (5.15) and (5.16). Plotting the theoretical prediction (5.15) as $(J^*/J_0)^{-2}$ vs B/B_T, as shown in the inset of Fig. 6.9(b), one finds a quasi linear dependence over a wide range of magnetic fields. Hence, one may approximate the critical current density for fixed temperatures as J^* ~ $(1+B/B_T)^{-1/2}$, consistent with the obtained scaling results according to (6.5) with an exponent close to $1/2$ for both samples. Furthermore, according to the definition (5.16) of the normalization current density J_0 it follows that J^* [given by (5.15)] depends directly on the quasi-particle scattering rate $1/\tau_{in}$ which in turn was shown to follow (6.4) in the high-temperature regime. Combining all three equations yields

$$J^* \propto f(T) = \left(1 - T/T_c\right)^{3/4} \cdot \exp\left[3.52 \frac{\left(1 - T/T_c\right)^{1/2}}{kT} \right] \tag{6.6}$$

As shown in Fig. 6.9(b), the function $f(T)$ can be similarly approximated by a power law with an exponent of $3/2$ for high temperatures $T \gtrsim 0.8\ T_c$ resulting in the observed linear dependence for both samples.

6.3 Angular Dependence of the Instability

As discussed extensively in Chap. 3, the high anisotropy of the $Bi_2Sr_2CaCu_2O_{8+\delta}$ system may cause an effective decoupling of CuO_2 planes and a two-dimensional behavior according to the model developed by Kes and Clem [Kes90, Clem91]. In this case only the c-axis component of a magnetic field applied to the superconductor will produce vortices, whereas the component parallel to the CuO_2 planes can penetrate completely. Thus a quantity Q which depends on the magnetic field applied parallel to the c-axis direction should exhibit an angular dependence of the form

$$Q(\mu_0 H, T, \theta) = Q(\mu_0 H \cos\theta, T, \theta = 0°) \qquad (6.7)$$

where Q may represent a specific parameter such as the critical current I^* or the entire CVC. As it was shown in the preceding section that both I^* and V^* depend on the magnetic-field strength applied parallel to the c axis, one will expect these critical parameters to scale with the angle according to (6.7) if the 2D model accurately describes the BSCCO system.

For fixed temperatures and magnetic fields CVC's were taken at different orientations θ of the magnetic-field direction relative to the c axis of the sample. An example of I–V isotherms for $\mu_0 H = 0.5$ T and T = 76 K at angles θ = 0, 20, 30, 40, 50, 60, 70, 80, 90 degrees is displayed in Fig. 6.10. The double logarithmic plot of these I–V curves in the inset manifests an alike regime of flux flow as in the case of $H \parallel c$ clearly preceding the instability. Evidently, the instability occurs for all values of θ, with the critical current increasing and the critical voltage decreasing as the magnetic field is tilted away from the c axis. Throughout the entire accessible measurement range of temperatures and magnetic fields this anisotropic behavior is conserved. The variation of I^* and V^* with θ strongly resembles their dependence on H (for $\theta = 0°$) and suggests a decrease of an effective magnetic field, i.e. a decrease of the component contributing to the dissipation mechanism. As a test of the model of Kes, Fig.

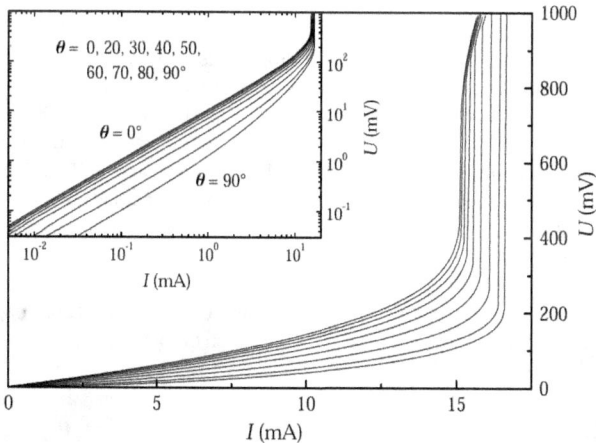

Figure 6.10: CVC's for fixed magnetic field $\mu_0 H$ = 0.5 T and temperature T = 76 K displaying the voltage instability at all angles θ = 0–90°.

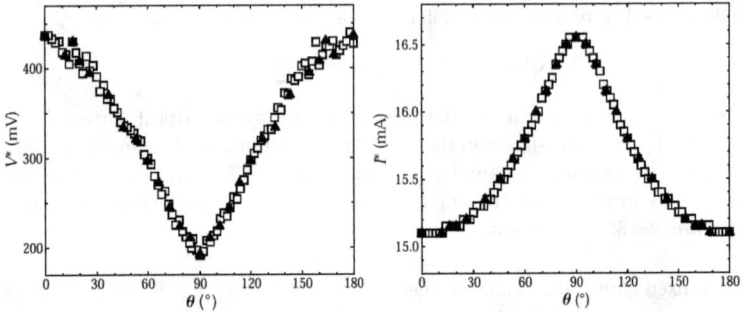

Figure 6.11: Angular dependence of the critical current density and voltage. Open symbols were measured at the respective angles; solid symbols were calculated according to (6.7) from data recorded at $\theta = 0°$.

6.11 presents the angular dependence of I^* and V^* at a fixed temperature $T = 76$ K. In this plot open squares indicate critical values extracted from I–V curves taken with a fixed magnetic field $\mu_0 H = 0.5$ T at the respective angles, $0° \leq \theta \leq 180°$. Solid triangles reflect values calculated from data taken at fixed angle $\theta = 0°$ ($H \parallel c$) and various magnetic fields $\mu_0 H$. The critical parameters I_1^* and V_1^* extracted from the latter measurements were then scaled according to (6.7) such that $\mu_0 H_i (\cos \theta_i)^{-1} = 0.5$ T. The agreement between both sets of data shows that for both I^* and V^* the two-dimensional behavior of the Kes model is indeed applicable.

Bearing in mind that, according to (5.14) and (5.16), J_0 and E_0 are independent of magnetic field, one will thus expect them to display no angular dependence either. With (6.7) incorporated into (5.12) and (5.15) this is tantamount to an identical relationship between E^* and J^* for all orientations. Correspondingly, in a simultaneous plot of E^* vs J^* extracted from CVC's one will expect all data for different magnetic fields and angles but a fixed temperature to fall onto a common curve, as shown in Fig. 6.12(a). Here, open circles indicate the data obtained at $\theta = 0°$ (cf. Fig. 6.5) in the entire accessible magnetic-field range at three temperatures, $T = 65$ K, 71 K, and 76 K; remaining symbols represent data taken at the same temperatures but different angles [as listed in Fig. 6.12(b)] and magnetic fields.[28] In both cases the experimental results agree well with the theoretical predictions of (5.12) and (5.15) plotted as solid lines for all three temperatures. Hence, one obtains the same normalization parameters E_0 and J_0 independent of the orientation of the magnetic field, as expected within the 2D model.

---

[28] Limited by excessive voltage increases at low B (which threaten to destroy the sample) and by the disappearance of the voltage jump at high B, the accessible magnetic-field range changes with θ and differs for all sets of data. Coinciding single data points in this plot indicate equal magnetic-field components along the c axis, $\mu_0 H \cos\theta$.

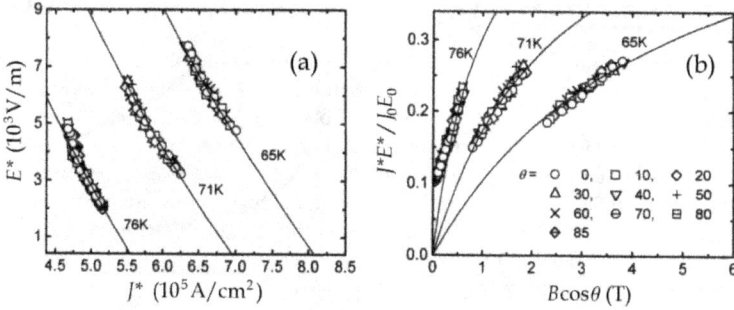

Figure 6.12: Relationship of the critical electric field, current density and magnetic field at various angles for three different temperatures. Using the magnetic-field component parallel to the c axis allows to describe all data according to the universal functions [solid lines, derived from (5.12) and (5.15)].

Following the approach of [Bezu92] to extract τ_{in} from B_T it is possible to investigate the anisotropy of the quasi-particle scattering rate from the characteristic fields for different orientations. If the data is plotted as J^*E^*/J_0E_0 vs $B\cos\theta$, as shown in Fig. 6.12(b), the collapse of the data onto a common curve for all magnetic fields and angles at a fixed temperature indicates that the dissipated power density of the instability in various field directions remains constant as long as the c-axis components of the magnetic field are equal. As a consequence, the characteristic field B_T also depends only on the component of B perpendicular to the CuO_2 planes, and $\tau_{in} \propto B_T = B_T(\theta=0°)/\cos\theta$.

The angular dependence of the upper critical field H_{c2} could also be suspected as the origin of the observed behavior. Both the quasi-2D model [Lawr70, Silv97] and the anisotropic 3D model [Tink96] predict a simple form of the anisotropy of the upper critical field $H_{c2}(\theta) \approx H_{c2}(\theta=0°)/\cos\theta$ for a high anisotropy parameter. It follows for the quasi-particle diffusion coefficient $D \propto 1/H_{c2} \propto \cos\theta$ [Bezu92] and, combined with (5.14), (5.16), and the observed angle independence of J_0 and E_0, one also obtains $\tau_{in} \propto 1/\cos\theta$. These models assume that the vortices consist of both longitudinal and transverse components of the magnetic field. However, Bitter decoration experiments have shown that vortices are formed only by the c-axis components of B [Boll91] and recent muon-spin-rotation measurements have provided microscopic evidence for the transparency of the CuO_2 planes of BSCCO to magnetic fields [Koss98].

6.4 Conclusions on the Vortex Instability in BSCCO

With transport measurements at high dissipation levels for the first time a voltage instability was detected in Bi₂Sr₂CaCu₂O₈₊δ, the second high-temperature superconductor (besides YBCO) found to exhibit this phenomenon. Regarding the extended free flux flow regime of the vortex dynamics in the BSCCO system, the temperature and magnetic-field dependent jumps in CVC's can be understood in terms of the Larkin-Ovchinnikov model of a vortex instability at high vortex flux-flow velocities. While possible origins of the discontinuities in the current dependence of the resistivity other than the LO mechanism can be safely excluded, the experimentally observed magnetic-field dependence of the critical vortex velocity v^* and the current density J^*, where the instability occurs, cannot be explained by the pure LO model. However, the extension to this model by Bezuglyj and Shklovskij, taking into account the effects of inevitable self heating of quasi-particles, gives a good quantitative description of the data in the entire temperature and magnetic-field range. An in-depth analysis of the current densities and electric fields of the instability points yielded the temperature dependent characteristic magnetic field B_T, separating the regimes where the mechanisms of quasi-particle heating ($B > B_T$) and of the actual LO instability ($B < B_T$) dominate. The determined quasi-particle diffusion length is consistent with the assumptions of the BS model, i.e. it exceeds the vortex core size and the intervortex distance. The extracted inelastic quasi-particle scattering rate $1/\tau_{in}$ exhibited a temperature dependence which could be explained quite precisely by the exponential dependence of the electron-electron interaction on the variation of the superconducting gap with T. The absolute values of $1/\tau_{in}$ compare well with experimental results published for the related YBCO system.

Finally, the angle dependence of the vortex instability was investigated. For all orientations of the magnetic field the observed results were well described by a quasi-2D model of effectively decoupled pancake vortices. This is consistent with the results on the magnetic-field dependent dimensionality from Chap. 3, where three-dimensional behavior was strictly confined to the regime of extremely low magnetic fields, $\mu_0 H < 0.1$ T, whereas the measurements of the instability experiments were carried out at higher fields. Furthermore, the activation energy and the vortex glass transition, which constituted the basis for the analysis of sample dimensionality in Chap. 3, both relate to vortex dynamics at low current densities where the effects of pinning play a fundamental role. While dimensionality for low J is thus related to interlayer coupling, the mechanism of the instability at high J—and hence its effective dimensionality—relies on dissipation processes occurring in the vortex. Obviously, the dissipation mechanisms relying on the excitation and relaxation of quasi-particles in the LO model occur in the superconducting CuO₂ planes and only the vortex components in these planes contribute, resulting in the observed quasi two-dimensional dependence of the effective magnetic field $\mu_0 H_{eff} = \mu_0 H \cos\theta$.

7 Correlation of Instability and Vortex Glass in YBa$_2$Cu$_3$O$_7$

As mentioned above, the voltage jumps ascribed to the Larkin-Ovchinnikov vortex instability had been observed in HTSC's initially in the YBa$_2$Cu$_3$O$_7$ system [Doet94, Xiao96]. In these analyses the inevitable influence of quasi-particle heating had not been taken into account, and the successful interpretation of the BSCCO data with the Bezuglyj-Shklovskij extension to the LO model [Bezu92] given in the previous chapter suggests the application of this theory to the YBCO system. Due to the different character of the phase diagrams in YBCO and BSCCO, the question arises to what extent the discrepant vortex dynamics at low flux-line velocities affect the nature of the vortex instability at high velocities. In YBCO the wide glass phase, extending to relatively high temperatures $T_g \gtrsim 0.9\ T_c$ within the accessible experimental range where the instability can be observed, allows to investigate the influence of the more three-dimensional character of the vortex system, the lack of an expanded FFF region, and the different associated pinning properties. With regard to the possibility of Joule heating the first part of this chapter deals with YBCO samples prepared on various substrate materials and the influence of the different heat conducting properties on the observed voltage instabilities. The following section concentrates on the BS analysis of the obtained data and the extracted quasi-particle parameters. The correlation between the low and high dissipative regimes, correlating the low electric-field glass phase with the appearance of discontinuous voltage jumps at higher E, is the focus of the third section.

7.1 Influence of Substrate Material

Possible influences of sample heating on the analysis of voltage instabilities have been noted [Xiao98a, Xiao99] and two approaches minimizing the effects of Joule heating in transport measurements of superconducting thin films were suggested. First, compared to quasi-DC measurements with pulse durations on the order of seconds, the use of rapid single pulses in the millisecond range [Doet94] will reduce

Table 7.1:   Parameters of four YBa₂Cu₃O₇ samples prepared on substrate materials with varying thermal conductivity k. l and w are the length and width of the measurement bridge, respectively. The transition temperatures (midpoint) $T_{c,mp}$ vary by less than 3 K and are quite comparable. However, the transition widths ΔT_c differ more strongly. A downset of the resistive transition at a reduced $T_{c,ds}$ reveals an inferior sample quality for two of the specimen, most likely due to inhomogeneities. As an additional indication of sample quality ρ_n, the normal state resistivity at 100 K, is given. ([a] [Toul70], [b] [Mich92])

Sample	Substrate	k (W/mK)	l (μm)	w (μm)	$T_{c,mp}$ (K)	ΔT_c (K)	$T_{c,ds}$ (K)	ρ_n
T301	SrTiO₃	~ 10[a]	100	10	89.0	2.2	87.0	180
Y001	SrTiO₃	~ 10[a]	500	50	91.7	3.1	86.3	136
Y026	LaAlO₃	18.6[b]	500	50	91.0	0.8	90.3	208
Y081	MgO	≥ 40[a]	100	10	88.1	1.8	87.0	143

the amount of heat deposited in the sample during measurement.[29] This short-pulse approach will be the focus of the subsequent chapter. Second, the heat conductivity of the substrate material may play an important role in an effective heat transport from film to bath and the use of MgO instead of SrTiO₃ with a considerably increased thermal conductivity has been suggested [Mars93, Doet94]. Thus, several YBa₂Cu₃O₇ films were prepared on different substrate materials. The samples T301, Y001, and Y015 on SrTiO₃, Y026 on LaAlO₃, and Y041 and Y081 on MgO were patterned and analyzed. For all samples the quality, including morphology, transition temperature T_c, and transition width ΔT_c, differed slightly. An overview of the relevant parameters is given in Tab. 7.1 for the four samples studied in depth.

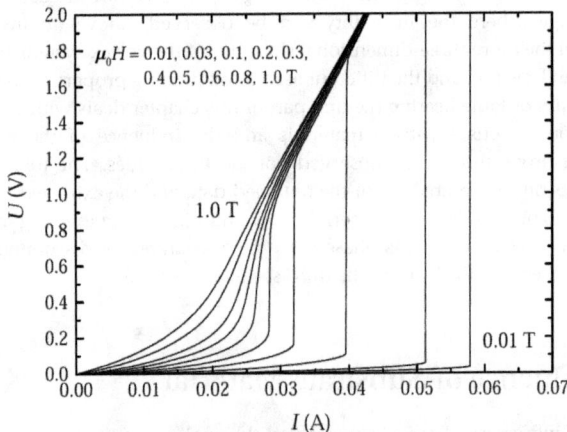

Figure 7.1: I–V curves of Y026 on a linear scale for fixed $T = 89$ K at magnetic fields $\mu_0 H = 0.01$–1.0 T.

---

[29] Undesired but avoidable excessive Joule heating of the crystal lattice should not be confused with inevitable quasi-particle heating according to the BS model.

In transport measurements at high dissipative levels the CVC's of all samples display voltage jumps, as shown in Fig. 7.1 for sample Y026 (LaAlO$_3$) at a fixed temperature. At low T (and low B) a discrete voltage jump occurs for a well defined I^* and V^*. If either temperature or magnetic field are raised I^* decreases while V^* increases. At sufficiently high B or T the instability (i.e. the discrete jump) disappears and only a continuous voltage increase remains visible. A direct quantitative comparison of the various samples is not conclusive due to the different normal resistivities, transition temperatures and widths. However, all samples exhibit the same qualitative dependence on B and T. Thus, while the substrate material may have an influence on the precise values of the critical current density J^* and electric field E^* even a considerably changed thermal conductivity leaves the character of the voltage instability itself unaltered.

7.2 LO and BS Quasi-Particle Scattering Rates

In light of this result, an analysis of the data was conducted for all of the above samples according to the extended LO model of a vortex instability which accounts for quasi-particle heating. The procedure, in analogy to Chap. 6, consisted of a fit of the recovered values of J^* and E^* (determined from CVC's taken at $75 \leq T \leq 84$ K and 0.1 T $\leq \mu_0 H \leq 6$ T) to the universal curves (6.2) and (6.3). The obtained results for T301

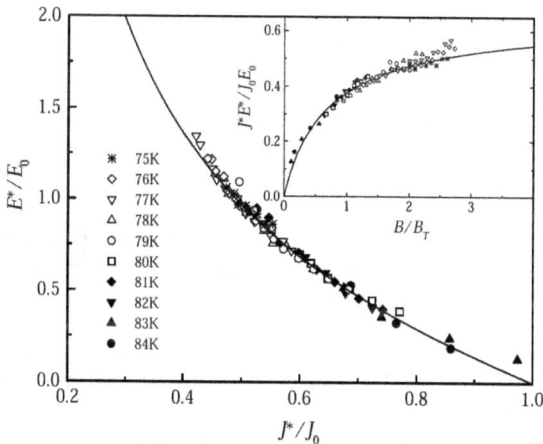

Figure 7.2: BS analysis of J^* and E^* for sample T301 (YBCO on SrTiO$_3$). Excellent agreement between experimental data (symbols) and theory (solid lines) is obtained for a wide range of temperatures, 75 $\leq T \leq$ 84 K, and magnetic fields, 0.1 T $\leq \mu_0 H \leq$ 6 T. The deviations of the experimental data from the theoretical prediction in the plot of J^*E^*/J_0E_0 in the inset are due to an overestimate of E^* arising from the widening of the instability region at high magnetic fields where the jumps begin to disappear.

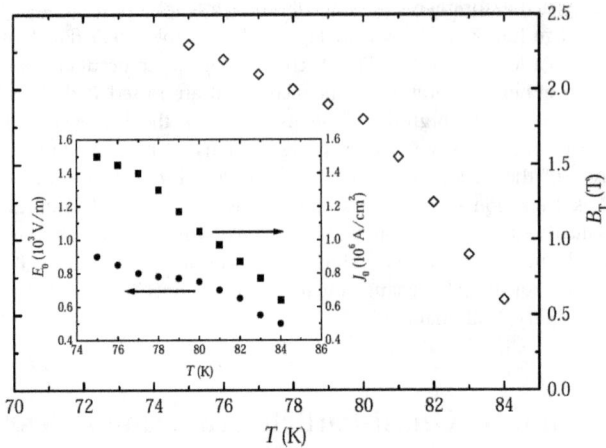

Figure 7.3: Characteristic magnetic field B_T and normalizing parameters E_0 and J_0 vs T for sample T301.

are presented in Fig. 7.2, where both plots of the experimental data, E^*/E_0 vs J^*/J_0 and J^*E^*/J_0E_0 vs B/B_T, are very well described by the theoretical predictions. At each temperature, data for larger magnetic fields at which the jump begins to disappear were excluded from this analysis: due to the widening of the jump region with an upward curvature and due to the limited current resolution the extracted values of E^* will likely be overestimated in these cases. This effect is still visible in the inset of Fig. 7.2, where the experimental data for J^*E^*/J_0E_0 begins to deviate systematically from the theoretical curve at large magnetic fields in the low-temperature range ($T <$ 80 K). The parameters J_0, E_0, and B_T for sample T301 are given in Fig. 7.3; the corresponding values for sample Y001 range from $J_0 = 6.0 \times 10^5$ A/cm^2, $E_0 = 540$ V/m, and $B_T = 2.6$ T at 81.0 K to $J_0 = 2.7 \times 10^5$ A/cm^2, $E_0 = 280$ V/m, and $B_T = 0.2$ T at 89.5 K.

As discussed in Chap. 6, from the fitting parameters one may directly obtain the heat-transfer coefficient h, the diffusion length l_e, and the inelastic quasi-particle scattering rate $1/\tau_{in}$. The calculated value of $h \approx 60$ W/cm^{2}K for T301 is considerably lower than those previously reported for YBCO $h \sim 10^2$–10^3 W/cm^{2}K [Nahu91, Xiao98b]. However, this calculated coefficient depends on the entire experimental setup—including the thermal contacts between substrate and copper sample holder and the thermal properties of the sample holder itself—as well as the thermal boundary resistance between the YBCO film and the SrTiO$_3$ substrate which may substantially reduce the effective value of h.[30] The diffusion length exceeds 100 nm for all temperatures and thus fulfills the conditions $l_e > \xi$ and $l_e > a_0$ such that the

---

[30] The Kapitza resistance may be of major importance in the process of heat transport from film to bath as was shown in an experiment based on the technique developed in Chap. 8 and conducted after the completion of this work [Bass99b].

quasi-particles can escape from the vortex cores and their distribution will be homogeneous throughout the superconductor (cf. Sec. 6.2.4). For sample Y001 $h \approx$ 55 W/cm²K and $l_e > 200$ nm was found in good agreement with the above results.

The temperature dependence of the inelastic quasi-particle scattering rate $1/\tau_{in}$ (Fig. 7.4) was determined from the simple LO theory (open circles) according to (5.9) with $1/\tau_{in}^{1/2} \propto v_{LO}^* = E^*/\mu_0 H$ and from the BS extension (solid circles) as described in the previous chapter. Due to the strong magnetic-field dependence of the critical electric field E^* the extraction of $1/\tau_{in}$ solely based on the LO theory has been difficult and resulted in a variation of $1/\tau_{LO}$ by more than an order of magnitude at a constant temperature if $\mu_0 H$ was changed from 0.5 T to 5 T [Xiao96]. Therefore, the values for $1/\tau_{LO}$ (open circles) were obtained by averaging over the data in the a magnetic-field range for each temperature, where E^*/B was approximately constant. Both qualitative behavior and absolute values of $1/\tau_{LO}$ (~ 10^{10} s⁻¹) agree well with the results of previous analyses based on the simple LO theory [Doet94, Xiao96]. The temperature dependence of the LO scattering rate not too close to T_c has been described by a power law following (6.4) but with a temperature independent gap parameter $\Delta = \Delta(0) \approx 2kT_c$ [Doet94]. In particular, the rapid suppression of $1/\tau_{LO}$ below T_c was thus interpreted as an indication of the absence of electron-phonon scattering.

Yet, in the framework of the BS extension to the LO model the resulting scattering rate—especially its temperature dependence—differs considerably from that of the simple LO model. At temperatures a few Kelvin below the superconducting transition $1/\tau_{BS}$ begins to saturate and seems to approach a constant value as T is further decreased. The reduced values of $1/\tau_{LO}$ are a result of electron overheating as treated in the BS model. Comparing the expected critical electric fields for both models

$$E_{BS}^*/E_{LO}^* = t^{1/4}(3t-1)^{1/2}/\sqrt{2} \tag{7.1}$$

one finds a ratio of 1 for $B \ll B_T$, where $t \approx 1$ and both scattering rates may be extracted with either model. However, in the limit of $B \gg B_T$ with $t \sim 1/3$ where quasi-particle heating is expected to dominate the mechanism of instability, the experimentally observed critical electric field E_{BS}^* will be far smaller than expected in the LO model and hence $1/\tau_{LO}$ will be underestimated (cf. Sec. 5.3). Now, as $B_T(T)$ increases comparably slowly with decreasing T and the data at lower temperatures must be taken at increasingly higher magnetic fields (otherwise the voltage jumps become too large and the sample is destroyed) the ratio B/B_T grows and the disparity between $1/\tau_{LO}$ and $1/\tau_{BS}$ increases with decreasing temperature.

The observed saturation of $1/\tau_{BS}$ suggests that electron-phonon scattering does, indeed, play a role in the scattering processes at temperatures below T_c and requires a more precise description of the temperature dependence of the scattering rate than given in [Doet94] for the LO analysis with pure electron-electron scattering and a temperature independent gap. Extending the ansatz of (6.4), where the temperature dependence of the gap was considered for electron-electron scattering in the case of

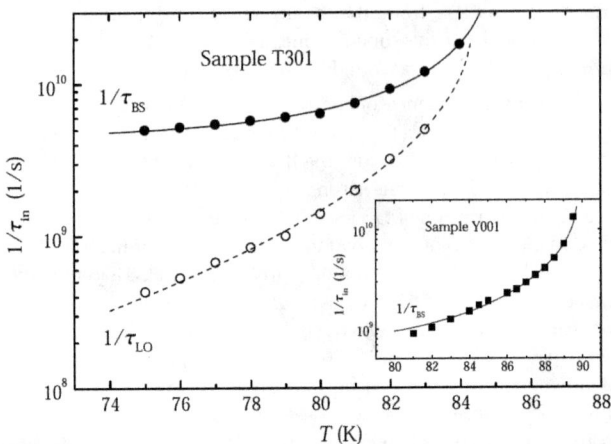

Figure 7.4: Temperature dependence of the inelastic quasi-particle scattering rate $1/\tau_{in}$ for sample T301 as determined from the simple LO theory (open circles) and from the BS extension (solid circles). Neglecting quasi-particle heating in the LO model results in a substantial underestimate of $1/\tau_{in}$ and hence of the resulting electron-phonon scattering contribution $1/\tau_{ep}$ at lower temperatures. The solid and dashed lines represent fits of (7.2) to $1/\tau_{BS}$ and $1/\tau_{LO}$, respectively, with $1/\tau_{ee} = 2.4(1) \times 10^{10}$ s⁻¹, $1/\tau_{ep} < 10^{6}$ s⁻¹, and $T_c = 84.8(2)$ K for $1/\tau_{LO}$; and $1/\tau_{ee} = 4.5(4) \times 10^{10}$ s⁻¹, $1/\tau_{ep} = 4.3(1) \times 10^{9}$ s⁻¹, and $T_c = 84.3(2)$ K for $1/\tau_{BS}$. In the inset the corresponding data and fit to $1/\tau_{BS}$ for sample Y001 is shown, where $1/\tau_{ee} = 1.9(2) \times 10^{10}$ s⁻¹, $1/\tau_{ep} = 0.61(7) \times 10^{9}$ s⁻¹, and $T_c = 89.7(2)$ K.

BSCCO (cf. Chap. 6 and [Xiao99]), by including a constant contribution $1/\tau_{ep}$ due to electron-phonon scattering one obtains

$$1/\tau_{in} = 1/\tau_{ee} \exp(-4\Delta/kT) + 1/\tau_{ep} \qquad (7.2)$$

The solid and dashed lines in Fig. 7.4 represent fits of equation (7.2) to the data of $1/\tau_{BS}$ and $1/\tau_{LO}$, respectively, yielding satisfactory agreement with the experimental data.[31] Besides the rates of electron-electron ($1/\tau_{ee}$) and of electron-phonon ($1/\tau_{ep}$) scattering in these cases also T_c itself had to be varied for the fitting procedure because the resistively determined $T_{c,mp}$ or $T_{c,ds}$ resulted in poor fits and unphysically high values of the scattering rates. In view of the considerably widened resistive transition the obtained value of $T_c \sim 85$ K for T301 does not seem unreasonable as the electron-electron scattering mechanism may be more sensitive to a widened transition than the resistive measurements. The extracted transition temperatures and scattering rates change somewhat if the upper limit of the temperature range used in the fit is extended to values close to T_c. However, temperatures too close to T_c should be excluded from such an analysis, especially for inhomogeneous samples, which can

---

[31] Supposing instead a temperature dependence of $1/\tau_{ep} \propto T$ results only in a minute variation of the parameters due to the small relative change in T.

Table 7.2: Heat-transfer coefficients h and quasi-particle scattering rates $1/\tau$ corresponding to electron-electron (ee) and electron-phonon (ep) interaction for YBCO samples on different substrate materials.

Sample	Substrate	h (W/cm²K)	$1/\tau_{ee}$ (s⁻¹)	$1/\tau_{ep}$ (s⁻¹)
T301	SrTiO$_3$	60	$4.5(4) \times 10^{10}$	$4.3(2) \times 10^{9}$
Y001	SrTiO$_3$	55	$1.9(2) \times 10^{11}$	$6.1(7) \times 10^{8}$
Y026	LaAlO$_3$	60	$1.3\text{–}5.0 \times 10^{11}$	–
Y081	MgO	200–500	$\sim 10^{11}\text{–}10^{12}$	–

be expected to exist in a non-uniform state with superconducting and normal regions coexisting at the same temperature.

In the case of the LO analysis the minute resulting electron-phonon scattering rate, $1/\tau_{ep} < 10^6$ s^{-1}, reduces (7.2) to the picture of quasi pure electron-electron scattering obtained previously [Doet94]. However, the fit to the data of the BS analysis yields a considerable contribution of electron-phonon scattering, $1/\tau_{ep} = 4.3 \times 10^9$ s^{-1} $\approx$ $0.1 \cdot (1/\tau_{ee})$ to the process of quasi-particle scattering in YBCO. These findings are corroborated by the experimental results of the BS analysis for sample Y001, shown with a corresponding fit in the inset of Fig. 7.4, where one observes the same saturation of $1/\tau_{in}$ for lower temperatures, a similar scattering rate on the order of 10^{11} s^{-1}, and also a reduced T_c obtained from the fit. Independent measurements of the microwave surface impedance of YBCO thin films [Gao93] and single crystals [Bonn93] also reported a less pronounced temperature dependence of the quasi-particle scattering rate similar to the BS analysis and in contrast to the steep suppression of $1/\tau_{in}$ of the pure LO analysis.

The analyses of samples Y026 and Y081 according to the BS model, particularly the fits to the universal curves, were less solid than for the above two samples but the qualitative behavior was preserved and the obtained results generally agree with the previous conclusions. For Y026 on LaAlO$_3$ a heat-transfer coefficient $h \approx 60$ W/cm²K and a total quasi-particle scattering rate $1/\tau_{in} \sim 3 \times 10^{11}$ s^{-1} was found; for Y081 on MgO with a higher thermal conductivity the heat-transfer coefficient $h \approx 200\text{–}500$ W/cm²K is clearly increased and $\tau_{in} \sim 10^{11}\text{–}10^{12}$ s^{-1} (Tab. 7.2). In both samples the observed saturation of $1/\tau_{in}$ at lower temperatures was less pronounced than for the YBCO films on SrTiO$_3$. Yet, this could be due to a more limited temperature range accessible in these measurements and the increased absolute value of $1/\tau_{in}$. Nevertheless, the deviating contributions of $1/\tau_{ee}$ and $1/\tau_{ep}$ to the total scattering rate along with the different characteristic magnetic fields B_T for the different samples and the sensitivity of (7.2) to an error in T_c raise the question of the influence of sample properties on the mechanism of the voltage instability and the extracted quasi-particle scattering rates. Obviously, great care must thus be exercised in the analysis of such high dissipation phenomena as a variety of other mechanisms may also come into play, for example the coexistence of a vortex instability and hotspots observed in a sample of inferior quality [Xiao98b]. In particular demonstrating the universality of the behavior for a variety of samples of different quality as well as on different substrates would be desirable.

7.3 Vortex Glass Phase Correlation

A feature independent of sample quality and substrate material and preserved in all analyzed YBCO samples is the correlation between the appearance of a discrete voltage jump and the vortex glass phase. It has been noted previously that such voltage jumps have appeared *also* below the glass temperature [Xiao96, Ruck97], however an actual correlation to the VG phase has not been reported. Such a dependence would be relevant for the mechanism of the instability itself, which could thus be related to the specific vortex pinning present in the sample, and due to the resulting correlation between the regimes of the low-dissipative vortex glass and the high-dissipative voltage instability.

As apparent in the CVC's of YBCO (Fig. 7.1) a discrete voltage jump at a well defined critical current density J^* is found only at sufficiently low temperatures and magnetic fields. If either T or B are increased the upturn in the I–V curve preceding the instability will become more pronounced and the *discontinuity* itself will shrink in size and finally disappear leaving an isotherm with a decisive but *continuous* voltage increase. The magnetic field at which this disappearance of the (discontinuous) jump takes place increases with decreasing temperature but the systematics remain unchanged at all experimentally accessible values of B and T.

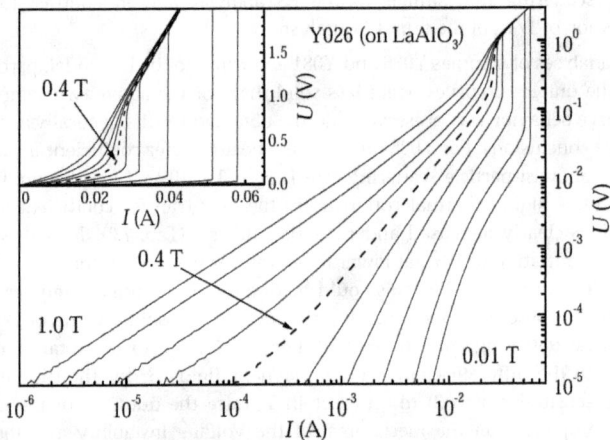

Figure 7.5: Double logarithmic plot of the CVC's of sample Y026 for $T = 89.0$ K and 0.01 T $\leq \mu_0 H \leq 1.0$ T. Downward curvature, observed in the I-V isotherms for $\mu_0 H < 0.4$ T, is characteristic of the vortex glass phase; upward curvature, visible for $\mu_0 H > 0.4$ T, generally indicates a fluid state. In the linear plot of the same data (inset) the transition from a discrete voltage jump at low magnetic fields $\mu_0 H \leq 0.4$ T to a continuous voltage increase for $\mu_0 H \gtrsim 0.4$ T can be clearly identified to occur at the glass transition around 0.4 T.

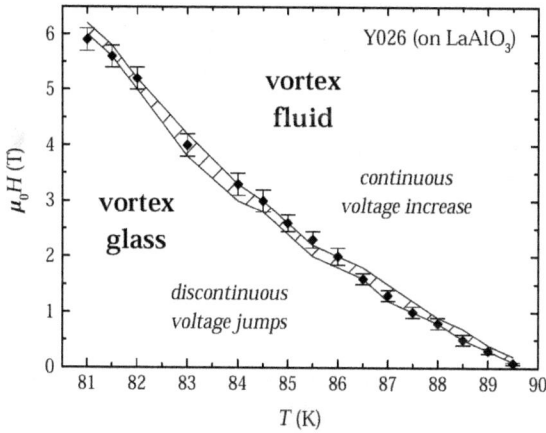

Figure 7.6: *B-T* phase diagram of sample Y026. The hatched region between the solid lines indicates the regime of the glass transition $B_g(T)$ as determined from the *I-V* isotherms. The symbols indicate the magnetic field B^* above which the discontinuous jump vanishes.

In the double logarithmic plot of the CVC's of sample Y026 for a fixed $T = 89.0$ K and various $\mu_0 H = 0.01$–1.0 T, shown in Fig. 7.5, a striking correspondence becomes apparent. For $\mu_0 H > 0.4$ T the isotherms exhibit an upward curvature at low current densities and electric fields attributed to a fluid phase, whereas below 0.4 T all isotherms possess a downward curvature characteristic of the vortex glass phase. The extended linear section in the low dissipative regime of the *I-V* curve at 0.4 T follows a power-law dependence of $E \propto J^r$ expected for CVC's at the glass transition [cf. (2.26)] separating the fluid and glass phases. In the high dissipative region where the voltage instability occurs the same isotherm at 0.4 T also separates the regimes of continuous voltage increase and discrete jumps. In the inset of Fig. 7.5, where the same CVC's are plotted on a linear scale, at magnetic fields below the glass line this discontinuous instability can be clearly distinguished from the gradual increase at fields above 0.4 T.[32]

From an analogous analysis of CVC's taken at temperatures between 81.0 and 89.5 K and in magnetic fields up to 6 T it is apparent that this correlation is not confined just to a limited range of magnetic fields and temperatures. The extracted *B-T* phase diagram for sample Y026 is presented in Fig. 7.6, including the glass line $B_g(T)$ as well as the temperature dependence of the magnetic field $B^*(T)$ at which the discontinuous jump vanishes. Obviously, the correlation between the vortex glass

---

[32] Unlike Chap. 4, where CVC's of different temperatures for a fixed magnetic field were analyzed, in this case the magnetic field is varied for a constant temperature. However, this does not alter the determination of the vortex glass transition in the *B-T* phase diagram.

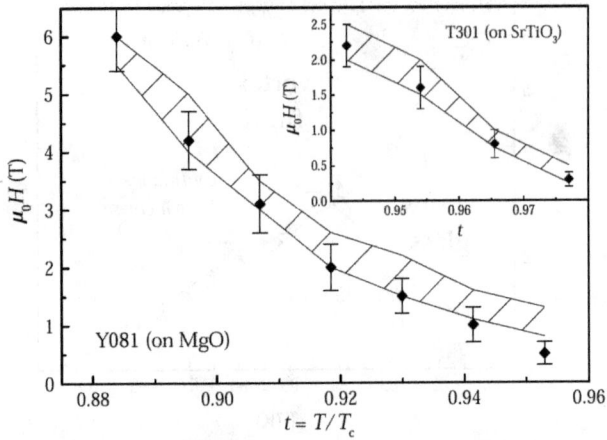

Figure 7.7:   B-T phase diagrams for YBCO samples Y081 on MgO and T301 on SrTiO₃ (inset) using reduced temperatures $t = T/T_c$ for comparison. The vortex glass transition occurs within the solid lines; the change of the instability from a voltage jump to a gradual increase is given by the solid symbols. The almost identical behavior for all three YBCO films of different sample quality and on different substrate materials strongly suggests that the correlation between the voltage instability and the vortex glass phase is a universal phenomenon of YBCO and possibly other superconductors with an extended VG phase.

phase and the discontinuous instability is preserved throughout the entire accessible phase diagram.[33]

Regarding the sensitivity of the BS analysis to sample quality found in the previous section, one might suspect that $B^*(T)$ may vary likewise, independent of the vortex glass line $B_g(T)$. However, the analysis of available data for samples T301 on SrTiO₃ and Y081 on MgO produced precisely the same correlation. Both phase diagrams also exhibit coinciding lines of the vortex glass and the instability transition, $B_g(T)$ and $B^*(T)$ respectively, over the entire measurement range (Fig. 7.7). This coincidence of B^* and B_g appears anything but coincidental. In fact, the shift of $B_g(T)$ to lower reduced temperatures and magnetic fields for sample Y081 but the preserved congruence with $B^*(T)$ strongly supports a direct correlation of the vortex glass phase and the voltage instability as a universal phenomenon in YBCO and possibly in other superconductors with extended VG phases.[34]

---

[33] An extensive VG analysis of the kind presented in Chap. 4 seems inappropriate in this case, as the comparatively short measurement bridge will result in wide range of suitable parameters for a vortex glass scaling and will not improve the results of the I–V glass isotherm analysis. Moreover, the dependence of the glass parameters on a high electric-field sensitivity is largely inconsequential for the determination of the glass temperature in this context, as T_g was observed to change by no more than about 1 K in the restricted and the extended measurement windows. Such a possible shift of the glass line is already included in the upper and lower limit for the VG transition in the phase diagram.

[34] The sole existence of a VG phase somewhere in the phase diagram is insufficient for the observed correlation, as in BSCCO the jumps clearly appear well above T_g despite a possible VG phase at low T (cf. Chap. 3).

7.4 Conclusions on the Vortex Instability in YBCO

Previous reports concerning the voltage instability in $YBa_2Cu_3O_7$ have mentioned the general importance of thermal effects but (in a quantitative analysis based on the Larkin-Ovchinnikov model) did not account for the influence of self heating of quasi-particles. By preparing several samples of YBCO films on various substrates with different heat conducting properties it was possible to show that the character of the observed instability remains unchanged and is essentially independent of substrate type and heat conductivity. The phenomenon in YBCO is thus due to an intrinsic effect of the superconductor rather than to Joule heating, as it would be the case for voltage jumps arising from hotspots or thermal runaway. On the other hand, in light of the quantitative agreement of the experimental data with the Bezuglyj-Shklovskij extension to the LO theory, it is apparent that quasi-particle heating plays an important role for the instability in YBCO, similar to the situation found in BSCCO. This process results from the thermalization of the power dissipated in the quasi-particle ensemble (corresponding to a deviation of the quasi-particle temperature from the phonon temperature of the crystal lattice) rather than from a limited heat transfer to the bath. The extracted quasi-particle scattering rates $1/\tau_{in}$ agree reasonably well with results of similar experiments. Moreover, a previously reported steep drop of $1/\tau_{in}$ at decreasing temperatures—attributed to pure electron-electron scattering—was shown to arise from the unjustified assumption of a simple LO mechanism which neglects quasi-particle heating. The analysis of the data based on the extended BS model yielded a saturation of $1/\tau_{in}$ at low temperatures instead, which is in better agreement with independent measurements of the microwave surface impedance of YBCO. This saturation is due to the steep decrease of the electron-electron scattering rate $1/\tau_{ee}$ but an approximately constant contribution (over temperature) of the electron-phonon scattering $1/\tau_{ep}$, which begins to dominate at low temperatures. Thus, it was possible to describe the temperature dependence of the total scattering rate by an extension of the approach used in Chap. 6 as $1/\tau_{in} = 1/\tau_{ee} + 1/\tau_{ep}$ yielding values of $1/\tau_{ee} \sim 10^{11}$ s^{-1} and $1/\tau_{ep} \sim 10^9$ s^{-1}.

One of the essential distinctions between YBCO and BSCCO has been discussed extensively in the first part of this work, Chaps. 3 and 4. The considerably lower anisotropy and the more three-dimensional character of the YBCO system give rise to an extended vortex glass phase observable at low electric fields. In the analysis of the glass transition in the low dissipative regime of the CVC's, complementing the investigation of the instability at high dissipation levels, a direct correlation of the vortex glass phase and the existence of the voltage jumps was found. A discrete voltage jump appeared only for temperatures and magnetic fields below the glass line $B_g(T)$ whereas for higher T and B only a continuous voltage increase could be detected. While the quality of the LO and BS analyses appears to depend to some extent on sample properties, the VG correlation was clearly proven for YBCO films on various substrate materials and of different qualities with the glass line and the 'jump line' coinciding for all samples.[35] This universality is of major importance for the field of vortex dynamics as it establishes a correlation for any given magnetic field and temperature between the dynamics at low and high vortex velocities, which in all other kinds of analyses so far have appeared distinct and unrelated. Yet, although the nature of the vortex dynamics is changed at high current densities, some underlying characteristic of the vortex system at low current densities is obviously preserved and leads to the emergence of an instability in the VG case only.

With regard to the magnetic-field dependence of the temperature $T^*(H)$, at which a voltage increase appears in resistive transitions $\rho(T)$ for fixed current densities, a similarity to the irreversibility line has been noted [Xiao96] and the angular anisotropy of J^* has been shown to follow that of the critical current density J_c [Xiao97]. This suggests a relation of vortex pinning to the mechanism of the voltage instability and thus differs from the basic assumption of the LO model, where the voltage jump arises simply from the velocity-dependent damping coefficient within the flux-flow regime and is basically independent of pinning. Alternative mechanisms which are related to a glass phase, such as self-organized criticality, have been implicated as possible origins of the instability. However, the quantitative agreement of the experimental data with theoretical predictions for these models is very limited [Xiao97], whereas the LO model (in the BS extension) yields a good description. Also, the SOC model supposes small intervortex distances, yet the instability is most pronounced for low B (i.e. large a_0) and disappears for high B.

A possible explanation of the VG-instability correlation combines the effects of vortex lattice depinning and the LO mechanism. In the VG state at low current densities all vortices will be essentially stationary resulting in negligible dissipation but at sufficiently high driving forces the entire ensemble is depinned. The system can then quickly reach the critical velocity and the LO mechanism leads to the jump-like voltage increase.[36] In the state of an interacting, partly pinned fluid above T_g, however, the damping force can exceed that of free flux flow as the flux-line motion is opposed not only by the viscous damping of the LO mechanism but also by the interaction of moving and stationary vortices. Thus, the flux lines possibly never reach a critical velocity above which the damping force decreases and no voltage jump occurs, yet the increasing current density still leads to a (continuous) voltage increase. For increasing temperatures, the system approaches FFF and the influence of pinning and vortex interaction will decrease but at the same time the system also

---

[35] Previously, the *independence* of the instability from the vortex glass phase of YBCO had been reported [Xiao96], however that analysis considered the general emergence of a voltage increase (which exist for $T > T_g$ as well) as opposed to an actually *discontinuous* instability (which is observed *only* for $T < T_g$). Also, for amorphous and multilayer films of the low-T_c superconductor Ta/Ge a broadening of the transition (resulting in several consecutive small voltage jumps) was observed near T_g but a discontinuity in the CVC's actually still appeared well above the glass temperature [Ruck97].

[36] Once depinned the glassy vortex system may actually return to a regular lattice formation as there is effectively no disorder. Yet, within the proposed model this possible recrystallization is not the cause of the instability.

approaches the normal state as the region of actual FFF in YBCO is restricted to a very small temperature regime directly below T_c. The larger effective damping coefficient in the fluid region could explain the absence of the instability in the fluid phase of YBCO as opposed to the FFF phase of BSCCO where the influence of pinning and vortex interaction is negligible, the damping coefficient depends only on viscous drag, and a vortex instability occurs.

Also, the secondary damping mechanism opposing vortex motion at velocities $v \gg v^*$ and preventing the sample from becoming fully normal directly above the jump, may play a decisive role for the emergence of a vortex instability. In the regime close below the glass line where the discontinuous jump is less pronounced the sample is clearly not fully normal directly above the jump although it exhibits a strongly increased resistivity. As of now, there exists no model describing the vortex dynamics for this characteristic gradual approach to the normal state conductivity appearing at these high flux-line velocities.[37] Further analyses will thus require closer scrutiny of the dynamic aspects of vortex motion at the point of instability and shortly thereafter. In particular, rapid current pulse measurements allowing the time-resolved investigation of the vortex instability are desirable.

---

[37] A possible origin could be the delocalization of quasi-particles due to the high vortex velocities and diffusion lengths (cf. Sec. 5.1). The shrinking of the vortices could thus be slowed down and the voltage increase limited.

8 Time-Resolved Measurements of the Vortex Instability

The broadening of the upturn which precedes the jump in I-V curves and complicates the extraction of I^* and V^* for high magnetic fields (Chap. 6), the disappearance of the discontinuous jump above the vortex glass line in YBCO (Chap. 7), and the fact that both YBCO and BSCCO samples do not always enter directly into the normal state above the jump, call for a more detailed investigation of the mechanism behind the vortex instability. Especially, the difference between the visible but gradual voltage increase above T_g and the discontinuous voltage jump below T_g is an interesting topic in light of the correlation between the instability and vortex glass discovered in this work as reported in the previous chapter. Though, neither the character of the vortex dynamics at the onset of the voltage jump nor a secondary damping mechanism at the upper end of the jump (preventing the complete breakdown of superconductivity by impeding further vortex acceleration) have yet been addressed in either theoretical or experimental works. Therefore, measurements with rapid current pulses of length $\delta t \ll 1$ s and time-resolved recording of the voltage during the entire pulse duration were conducted, complementing the quasi-DC measurements of Chaps. 6 and 7 with current pulses of $\delta t = 1$ s and the voltage measured at the end of the pulse only. In the first section of this chapter the setup developed to meet the requirements of such experiments is described. A brief investigation of heating effects for short current pulses in YBCO samples on various substrates follows. The central aspect of the chapter is the analysis of the time dependence of the vortex instability and its correlation with the glass phase first observed in the quasi-DC measurements. As this method could be successfully employed only at a very late stage of the present work these measurements were intended primarily to investigate the potential of this method itself and general qualitative time dependent characteristics of the instability.

8.1 Experimental Setup

Aside from the current supply and the voltage measurement, the setup of these experiments, depicted schematically in Fig. 8.1, was analogous to that of the quasi-DC measurements described in the previous chapters. Replacing the commercial current source, which could only supply rectangular current pulses with a duration in excess of 10 ms, a transistor current source was developed and adapted to the requirements of short current pulses for instability experiments (for details see [WagM98]). As the output current of this source is controlled by an input voltage, current pulses of arbitrary waveforms with durations in the range of microseconds can be obtained by using a suitable frequency generator with programmable signal shape. The source is able to supply currents up to about 200 mA with a response of $\sim 10^4$ A/s corresponding to a minimum rise time of about 20 μs for the maximum current amplitude. For the time dependence of the actual output current $I(t)$ the voltage input signal is not sufficiently accurate and recording the voltage $U_I(t)$ across a reference resistance R_r is advisable. Two identical differential amplifiers are used to measure the output current $[I(t) = U_I(t)/R_r]$ and the voltage across the sample $[U_s(t)]$ at any given time. Using identical elements for the recording of both of these signals is vital, as an unknown possible time delay in the processing of each signal can thus be compensated. Both signals are recorded simultaneously by a digitizing oscilloscope with a sampling rate of 150 MHz, replacing the nanovoltmeter of the quasi-DC measurements. Hence, the data allows to investigate the time evolution of the voltage U_s vs t as well as the shape of the I-V curves by plotting U_s vs $U_I \propto I$. All samples analyzed with this setup—including the YBCO films Y001 on SrTiO$_3$, Y026 on LaAlO$_3$, Y081 on MgO, and a BSCCO film Z135 on SrTiO$_3$—have been characterized in the previous chapters.

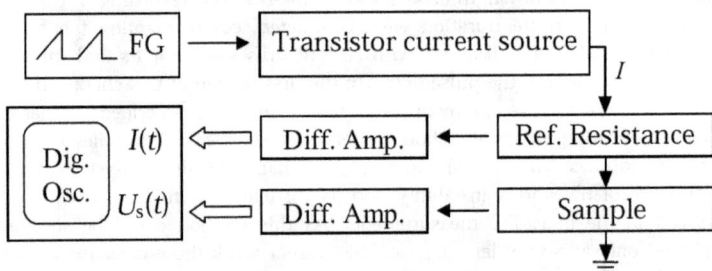

Figure 8.1: Schematic experimental setup for time-resolved measurements of I-V curves with short current pulses of arbitrary pulse form. The current applied to the sample is defined by the output voltage of the frequency generator (FG). Both sample voltage and current (across a reference resistance) are measured by identical differential amplifiers and recorded with a digitizing oscilloscope.

8.2 Heating Effects for Rapid Current Pulses

The possibility of varying not only temperature and magnetic field but also the pulse time parameters allows to investigate the thermal effects of dissipation at high current densities more closely than in the quasi-DC measurements. A very suitable approach is the comparison of CVC's for various pulse repetition rates. With the outlined setup it is possible to send current pulses of duration δt through the sample at well defined intervals Δt between the start of consecutive pulses, illustrated in Fig. 8.2. For all the experiments presented here a sawtooth pulse shape was used. Obviously, any amount of dissipation in the sample will lead to Joule heating and (due to the finite heat conductivity) to a rise of the sample temperature T_s above the bath temperature T_b. In the case of separate pulses repeated at a rate $1/\Delta t$ the deviation of T_s from T_b will be determined primarily by the heat deposited during the previous pulse and the interval time Δt, available for reapproaching thermal equilibrium with substrate and bath after Joule heating has ended. If, for a given pulse duration δt and maximum current amplitude I_m, the heat deposited during an individual pulse is sufficient to raise the sample temperature (on the order of 1 K or more), a change in the pulse interval time Δt will result in a visible change of the I–V curves. A shorter cooling time will lead to a higher sample temperature at the beginning of the next pulse and the I–V curve will be shifted as if taken at a higher temperature. The minimum cooling time above which no change in the I–V curves is detected will give an indication of the heat-transfer capabilities of the specific system.

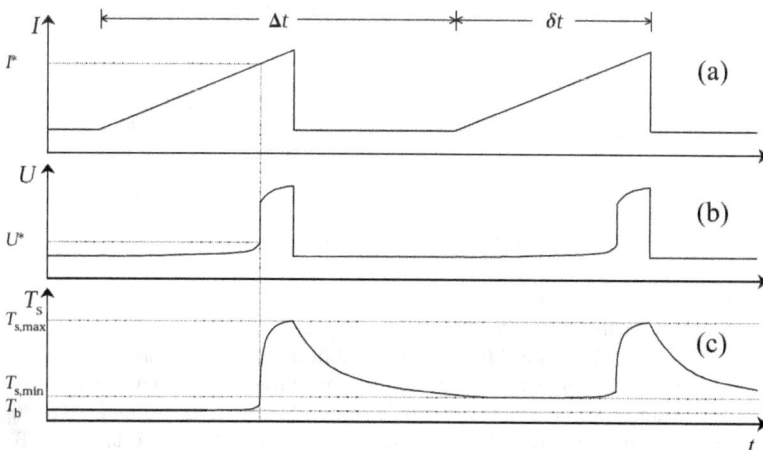

Figure 8.2: Schematic time evolution of (a) the current amplitude, (b) the voltage signal, and (c) the sample temperature for the initial two sawtooth pulses of length δt with a repetition rate $1/\Delta t$. The resulting deviation of the sample temperature T_s from the bath temperature T_b at the beginning of each pulse depends essentially on the heat deposited in the sample during the last pulse (i.e. $T_{s,max}$ at the end of a pulse) and the interval for cooling between pulses.

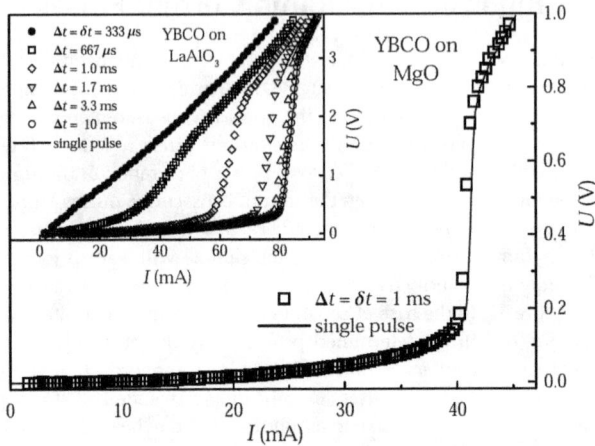

Figure 8.3: *I–V* curves for YBCO on MgO (Y081, $\mu_0 H$ = 0.3 T, T_b = 83.5 K) and LaAlO₃ (Y026, $\mu_0 H$ = 0.3 T, T_b = 88 K, inset) at fixed pulse times δt and various pulse repetition rates $1/\Delta t$ as indicated. The solid lines represent *I–V* curves obtained for a single current pulse, i.e. without any previous heating; symbols indicate data taken with consecutive pulses. For short pulse intervals of 1 ms and below (i.e. high repetition rates) in sample Y026 from the shift in the *I–V* curves a strong increase of sample temperature above bath temperature can be inferred, whereas for cooling times of 10 ms between pulses this shift vanishes (open circles) thus indicating thermal equilibrium of sample and bath. In sample Y081 the heat-transfer capability of the system is improved such that *I–V* curves taken with continuous pulses (open squares) still coincide with the single-pulse line, where $T_s = T_b$.

For sample Y026 (YBCO on LaAlO₃) this effect is clearly visible in *I–V* curves taken at a fixed (bath) temperature T_b = 88 K, magnetic field $\mu_0 H$ = 0.3 T, pulse duration δt = 333 μs and different interval times Δt (inset of Fig. 8.3). The obtained CVC's resemble those taken at different (bath) temperatures, as shown for example in Chaps. 6 and 7. Evidently for continuous pulses $\Delta t = \delta t$ = 333 μs the sample temperature is notably increased such that the *I–V* curve appears normal conducting even at low current densities. If Δt is increased the temperature difference between sample and bath is reduced and, for sufficiently long cooling times $\Delta t \sim$ 10 ms, the *I–V* curve coincides with the curve for a single pulse (where $T_{s,min} = T_b$). The situation for sample Y081 (YBCO on MgO) is quite different. Due to the enhanced heat conductivity of the substrate even for continuous pulses without *any* cooling interval (open squares) the curve can hardly be distinguished from the isotherm of a single pulse (solid line, Fig. 8.3). Thus, if there is no significant increase in T_s of YBCO samples on MgO even for continuous high-current pulses one may safely assume that—especially for single pulses—the sample (phonon) temperature will remain constant at the bath temperature.[38] Although for substrates of LaAlO$_3$ (and SrTiO$_3$, which displays a similar behavior) T_s is increased for larger repetition rates, this is essentially a result of the high dissipation arising in the state that the sample assumes *after* the instability has occurred but not of the low dissipation state corresponding to the *I–V* curve below the jump. Hence, even for LaAlO$_3$ and SrTiO$_3$ substrates a single pulse (without previous sample heating) will not lead to a substantial rise in T_s until after the instability has occurred. This is particularly important for the analysis of the quasi-DC measurements of the previous chapters as it corroborates that the voltage instability is not caused by thermal processes and that the actual sample temperature before the jump occurs will be close to the measured bath temperature.

8.3 Time Dependence of the Voltage Instability

For the analysis of the time dependence of the voltage instability it is essential to be able to exclude heating effects. Only if there are no time or current density dependent contributions of such effects to the voltage across the sample it is possible to isolate the actual vortex dynamics from the CVC's. While Joule heating below the onset of the jump is negligible even for LaAlO$_3$ and SrTiO$_3$ substrates as shown above, nevertheless, in order to minimize any possible thermal influences before and during the instability, the following experiments were conducted on MgO substrates. The experimental results for both available samples with YBCO on MgO agree; in the following the data for Y081 will be presented, which was also investigated extensively in quasi-DC measurements. Figure 8.4 displays CVC's obtained from sample Y081 at $\mu_0 H$ = 3.0 T and T = 72.0–87.0 K. Each *I–V* curve consists of simultaneously measured $I(t)$ and $V(t)$ data for a single sawtooth pulse of length δt = 1 ms. As this short duration of the pulse strongly reduces the total amount of heat deposited in the sample compared to the quasi-DC measurements, it is possible to extend the measurement range to lower magnetic fields and temperatures without destroying the sample despite the larger currents and voltages involved. The character of the instability in the CVC's remains unchanged in the time-resolved measurements. While at high temperatures $T \gtrsim 80$ K the voltage increase with current appears gradual, at $T \leq 79$ K this increase appears as a sharp jump.

However, within the time resolution of 2 μs per data point this jump—although sharp—is certainly not instantaneous (as it appeared in the quasi-DC measurements) even at the lowest temperatures, where the voltage increase during the instability is most pronounced. Generally, this 'extended' jump could be ascribed to two different causes: a current dependence as well as a time evolution of the voltage increase. The former mechanism could be identified with a current interval $I^* < I < I^* + \Delta I$ over

---

[38] If the heat deposited during a single pulse is increased considerably (for instance, if the *I–V* curve extends far beyond the voltage jump) one will find a shift of the isotherms even on MgO substrates but only for very high repetition rates.

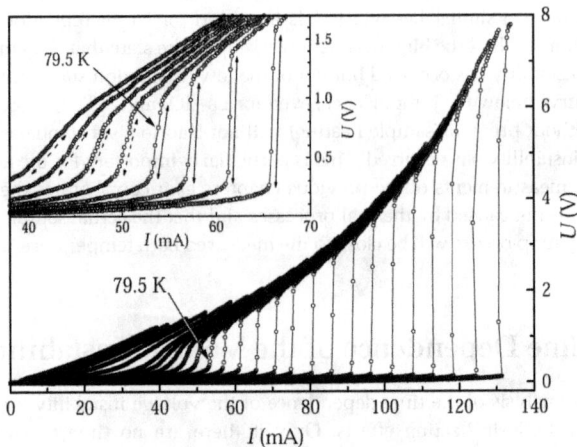

Figure 8.4: High current density CVC's of sample Y081 taken with the short-pulse setup at $\mu_0 H = 3.0$ T and $T = 72.0$–87.0 K. All curves were obtained with single pulses of length $\delta t = 1$ ms, thus excluding any sample heating preceding the voltage instability. A single data point corresponds to a time interval of $2\,\mu s$. The arrows in the inset illustrate the definition of the rise time Δt_R.

which the voltage increase extends. In principle, such a dependence could have been obscured in the quasi-DC measurements by the limited resolution (step width) of the current source. Yet, closer inspection reveals that the 'width' of the jumps in Fig. 8.4 (~ 1 mA) clearly exceeds this resolution of 0.1 mA. The latter explanation would relate to a rise time Δt_R which the instability needs to fully develop after the initial voltage increase.[39] Naturally, for Δt_R on the order of microseconds this widening of the voltage jumps must have been concealed in the quasi-DC measurements, where the voltage was recorded 1 s after the current had reached its maximum.

I–V curves in Fig. 8.5, taken at fixed magnetic field and temperature but different pulse times δt, exhibit apparently sharp jumps only for pulse times $\delta t \geq 1$ ms, whereas for a shorter pulse duration the curves are shifted to higher currents and the enhanced time resolution clearly reveals extended voltage increases. The values of Δt_R, given in the inset of Fig. 8.5, decrease from $400\,\mu s$ down to below $4\,\mu s$ as the pulse lengths are changed from 100 ms to 33 μs. Whereas the notion of the instability being extended over a *current* (rather than time) interval could possibly explain the apparently sharp increase for long δt as a result of Joule heating and the increased resistivity during the instability, it fails to account for the shift of the onset of the

---

[39] This rise time can most conveniently be defined as the time corresponding to the visibly pronounced increase of the voltage in the I–V curves, as indicated in the inset of Fig. 8.4. A more rigorous definition would require a theoretical description of the I–V curves in the high-dissipative regime, which is still lacking.

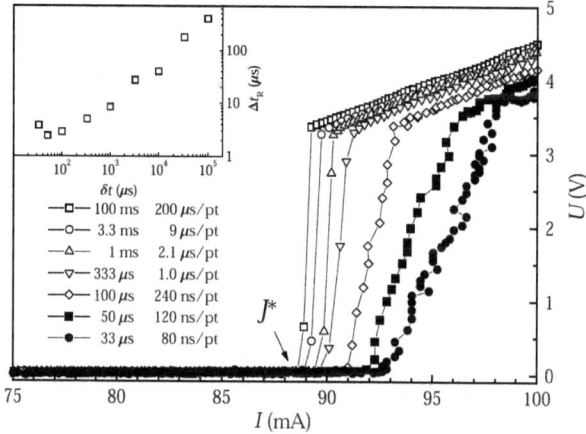

Figure 8.5: *I–V* curves of Y081 for fixed $T = 81$ K and $\mu_0 H = 0.1$ T but different pulse times $33\,\mu s \leq \delta t \leq 100$ ms. For larger $\delta t \geq 1$ ms (and a time resolution $> 1\,\mu s$ per data point) the jump appears sharp. However, at shorter $\delta t < 1$ ms and an improved time resolution $\leq 1\,\mu s$ per data point) it becomes evident that the voltage instability develops over an extended period of time. Given in the inset are the resulting rise times for the different pulse lengths.

voltage increase visible for shorter pulse times: for $\delta t \geq 1$ ms the voltage already reaches the upper segment of the *I–V* curve at currents $I \leq 90$ mA, but for $\delta t \leq 0.1$ ms the voltage increase does not *begin* until $I > 90$ mA. As shown in the previous section, there is certainly no heating below the onset of the instability for YBCO on MgO.

Thus, the shift in the onset of the voltage increase for short pulses (i.e. a fast current increase) can only be the result of a time delay in the response of the vortex system. Supposing that the instability develops over an extended period of time instead of quasi-instantaneously, one can qualitatively explain the behavior of the CVC's in Fig. 8.5. After the critical current density is reached, the voltage instability requires a certain time Δt_s to set in. For long pulses $\delta t > 1$ ms, Δt_s is smaller than the time resolution ($\sim 10\,\mu s$) and thus 'hidden' in the first data point. If pulse times are shortened to $\delta t < 0.1$ ms and the time resolution improved to $< 1\,\mu s$, Δt_s becomes visible as the shift in the *I–V* curves (hence indicating an offset in time rather than an actual increase of J^*). As the voltage across the sample is essentially determined by the vortex dynamics it is understandable that the entire voltage increase itself should also occur over an extended period of time, manifested as the finite rise time Δt_R. The exact value of Δt_R should depend on the particular conditions in the vicinity of the instability. In fact, the observed decrease of Δt_R for small δt may be an indirect consequence of the aforementioned time delay of the onset of the instability after J^* is reached. The resulting shift of the *I–V* jump to higher J is notably stronger for shorter pulse times and, in these cases, leads to an increase in the current density at the onset of the voltage jump on the order of a few percent. While this increase may seem

113

Table 8.1: The onset time Δt_s, which passes after J^* is reached and before the voltage increase begins, the rise time Δt_R, corresponding to the period of actual voltage increase, and the total time $\Delta t_t = \Delta t_s + \Delta t_R$, which is needed for the instability to fully develop, for different pulse times δt and time resolutions (i.e. time per data point).

δt (ms)	Resolution (μs)	Δt_s (μs)	Δt_R (μs)	Δt_t (μs)
100	200	–	400	400
33.3	90	–	180	180
10.0	20	20	40	60
3.33	9	9	27	36
1.00	2.1	6.3	8.4	14.7
0.33	1.0	3.0	5.0	8.0
0.10	0.24	1.9	2.9	4.8
0.05	0.12	1.7	2.4	4.1
0.03	0.08	1.6	3.8	5.4

small compared to J, the rate at which the instability develops may depend sensitively on the excess current density $\Delta J = J - J^*$ rather than J^* itself. Thus, even a minute increase of J above J^* may have important consequences for the actual vortex dynamics, leading to shorter rise times Δt_R for shorter pulses and increased current densities. The values for the onset time Δt_s, the rise time Δt_R, and the total time $\Delta t_t = \Delta t_s + \Delta t_R$, which is needed for the instability to fully develop once J^* is reached, are compiled in Tab. 8.1.[40]

Determining the change of the rise time Δt_R with temperature (i.e. with varying size of the voltage jump) for a fixed pulse length $\delta t = 1$ ms from the CVC's of Fig. 8.4, one finds a constant value of $\Delta t_R = 10$ μs (Fig. 8.6) below $T^*(\mu_0 H = 3 \text{ T}) \approx 79$ K where a sharp jump exists in the CVC's. The change in the I-V curves to a gradual increase above T^* is accompanied by a sudden increase of Δt_R and beyond this temperature it becomes increasingly difficult to determine Δt_R consistently (cf. inset of Fig. 8.4). Yet, despite this larger uncertainty in Δt_R for higher temperatures, at T^* there exists a clear change in the character of the voltage increase, which appears for all magnetic fields investigated between 0.1 and 6 T.

Although T^* decreases for larger fields (as could be expected from the quasi-DC measurements) below T^* the rise time always assumes the same value of 10 μs independent of magnetic field and temperature. As shown in the inset of Fig. 8.6 for five values of $\mu_0 H$, all curves coincide at $\Delta t_R = 10$ μs below T^* and display a similarly strong temperature dependence of Δt_R above T^*.

---

[40] As far as the character of the instability itself is concerned sample Z135, the only BSCCO sample that could be investigated by this method, exhibited the same qualitative behavior as the YBCO films. Quasi-instantaneous jumps recorded at a low time resolution could be resolved as extending over several microseconds. Although an extensive analysis of the time dependent aspect of the instability in BSCCO remains to be conducted, the extended rise times appear to be a general phenomenon of the instability in type II superconductors rather than a unique feature of the YBCO system.

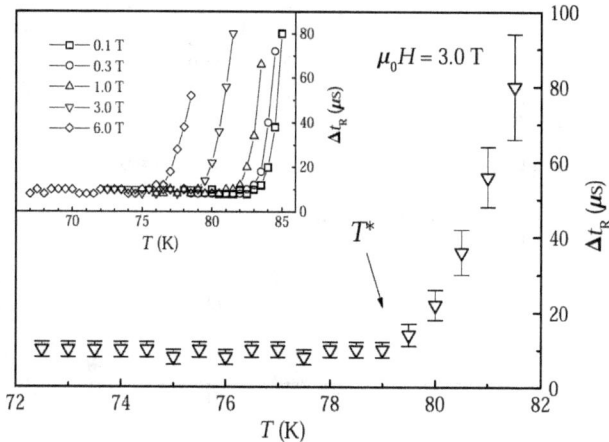

Figure 8.6:   Temperature dependence of the rise time $\Delta t_R(T)$ in sample Y081 as determined from the CVC's for $\mu_0 H = 3.0$ T and a pulse duration $\delta t = 1$ ms. Below $T^* = 79$ K, for all clearly recognizable voltage jumps, the rise time is constant at $10\,\mu$s. Above T^*, where the voltage increase becomes more gradual, the rise time grows considerably. This behavior is preserved at all magnetic fields investigated ($\mu_0 H = 0.1, 0.3, 1, 3, 6$ T), as shown in the inset, where T^* decreases with increasing H but the rise time for all sharp jumps is independent of T and H.

The sudden upturn of Δt_R is due to the change in the current-voltage dependence from an instability below T^* to a gradual (stable) increase above T^*. At low temperatures, the steady state of the vortex dynamics corresponding to a given current density exhibits a discontinuity at the critical value J^*: the massive change in the average vortex velocity corresponding to the increase in E cannot occur instantaneously when J is increased slightly above J^* and the system requires a period of time on the order of microseconds to assume its steady state. In the experiment this appears as the finite rise time found for all vortex jumps. However, for higher temperatures $T > T^*$ there is no discontinuity in $E(J)$ and hence the necessary change in vortex velocity corresponding to the increased current density can always be attained in time intervals below the resolution of the experiment. Thus, to be precise, there exists no genuine rise time in this case as the steady state corresponding to a given current density is assumed quasi-instantaneously. As a consequence CVC's above T^* should be time independent which is indeed observed in the essentially identical I-V curves for high T (and B) in the quasi-DC and short-pulse measurements (cf. Figs. 7.1 and 8.4). Still, the more gradual increase of E with J at higher T

115

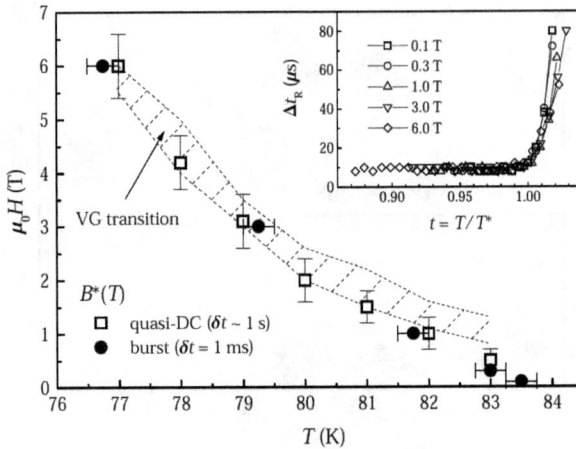

Figure 8.7: Phase diagram of Y081 (cf. Figure 7.7) with the VG transition (inside the hatched region), the jump limit $B^*(T)$ from quasi-DC measurements with $\delta t = 1$ s (open squares) and from pulse measurements with $\delta t = 1$ ms (solid circles). The determined jump limits agree quite precisely for both methods confirming the correlation between the instability and the VG phase. The inset shows the identical behavior of the rise time Δt_R for all magnetic fields plotted against the reduced temperature $t = T/T^*$.

will show up as an apparent increase of Δt_R, which thus allows to determine systematically the temperature boundary of the voltage instability.[41]

In Sec. 7.3 the identity of this boundary $T^*(B)$ [or equivalently $B^*(T)$] with the vortex glass transition $T_g(B)$ was reported for several YBCO samples, including Y081 on MgO inspected in these short-pulse experiments. Considering the time dependent nature of the voltage instability as well as the question of thermal effects in quasi-DC measurements, it is worthwhile to analyze once more the voltage instability in relation to the low dissipative vortex state. The limited sensitivity range of the experimental setup, in particular the oscilloscope and the non-linear current output of the source at lowest current densities, obstruct the conclusive analysis of the low voltage part of the time-resolved CVC's. In the quasi-DC measurements, however, the same measurement bridge of sample Y081 was investigated extensively yielding the B-T phase diagram with the vortex glass phase and the boundary of the voltage instability for long current pulses. Using T^* indicating the last appearance of a voltage instability for increasing temperatures as determined from Figure 8.6, one

---

[41] With regard to the apparently limited rate at which the vortex system can adjust to an increased driving force, it could be interesting to investigate the maximum response, i.e. the maximum possible acceleration of vortices at *different* dissipation levels. The use of current pulses of low maximum amplitude but extremely short duration (with a sufficiently high dI/dt) could allow this kind of investigation in the framework of low dissipation vortex dynamics as well.

finds excellent agreement with the results from the quasi-DC experiments and the vortex glass line, as shown in Figure 8.7. Also, the curves of the rise times Δt_R, if plotted against a reduced temperature $t = T/T^*$, coincide in the entire temperature range for all values of the magnetic field, thus giving further proof of the universality of the correlation of the instability and the vortex glass phase.

8.4 Conclusions from Time-Resolved Measurements

An experimental setup for rapid current pulses at high dissipation levels was developed and successfully employed in the measurement of current voltage characteristics and the recording of the time dependence of the voltage instability in high-T_c superconductors. Through variation of pulse parameters it was possible to investigate the thermal influence of the substrate material. Whereas LaAlO$_3$ and SrTiO$_3$ exhibited visible increases of the sample temperature due to Joule heating for high dissipation levels and short cooling intervals between pulses, in the case of MgO basically no change in the I–V curves was detected even for continuous current pulses. However, for cooling intervals on the order of 0.1–1 s none of the substrates displayed a notable temperature increase below the onset of the voltage instability. In the first place, these observations pertain to the conclusions drawn for the quasi-DC measurements of the previous chapters. The voltage instability does not originate from Joule heating and the sample (phonon) temperature will not begin to deviate substantially from the bath temperature until after the instability has occurred and the dissipation level is increased. At most, only a slight error in the determined sample temperature ($\delta T \lesssim 1$ K) will result for the extracted instability parameters $J^*(T)$ and $E^*(T)$, thus affirming the extracted quasi-particle scattering rates of the BS analysis as well as the identity of the instability line $B^*(T)$ and the glass line $B_g(T)$. Secondly, the absence of any thermal effects for MgO substrates is an important prerequisite for the investigation of the time dependence of the instability itself — a sample temperature increasing with time (particularly after the onset of the voltage jump) would obscure the actual time dependence of the voltage arising from the instability of the vortex dynamics.

In the analysis of the rise time Δt_R, i.e. the time interval of the voltage increase in the CVC's, it became apparent that there exist two regimes of different vortex dynamics in the B-T phase diagram. At sufficiently low magnetic fields or temperatures [below the instability line $B^*(T)$] a sharp, jump-like voltage increase occurs in the I–V curves at a critical current density; at the upper end of the voltage jump the sample is essentially in the normal state and the I–V curve reassumes a continuous behavior. A rise time of this voltage jump on the order of microseconds was detected and identified as a feature of the physical mechanism and not an experimental artifact since Δt_R varies as the pulse duration δt is changed. Yet, for a fixed pulse duration the jump-like voltage increase is characterized by a constant rise time at all magnetic fields and temperatures. At $B^*(T)$ the jump vanishes and for higher magnetic fields or temperatures [above the instability line $B^*(T)$] only a gradual voltage increase appears. This behavior can be understood in terms of a discontinuity existing below $B^*(T)$ in the current dependence of the voltage (relating to the vortex response to the driving force) such that the vortex system requires a finite amount of time after the critical current density is reached before it can assume the new dynamic state corresponding to the increased driving force. Above $B^*(T)$ the current dependence of the voltage is continuous and, strictly speaking, without an instability there exists no measurable rise time as the system assumes its final dynamic state quasi-instantaneously. However, with increasing B or T the interval of current density corresponding to the (now gradual) voltage rise substantially widens and the time needed to transverse this interval grows (for a fixed pulse duration) resulting in the apparently increasing rise time.

Values for the instability line $B^*(T)$ extracted from these time-resolved measurements revealed the same identity with the glass line $B_g(T)$ reported in Chap. 7 from quasi-DC measurements, thus fully confirming the correlation between the low-dissipative vortex glass phase and the high-dissipative vortex instability. The change in the CVC's from a gradual voltage increase above $B^*(T)$ to a discontinuous jump below is not due to a limited resolution of the current source but it is an intrinsic feature of the vortex dynamics, which change behavior both at low and high vortex velocities at the glass transition. Yet, while these measurements nicely demonstrate the predicted existence of the vortex instability, the question remains as to what causes the unpredicted and heretofore unknown finite response time of the vortex system on the order of microseconds.[42] Obviously, some process prevents the vortex dynamics from adapting arbitrarily fast to the changed conditions and several different mechanisms could be conceived as limiting the response of the vortex system.

Considering the interdependence of the vortex velocity and the accelerating net force $f_n(v) = f_\eta(v) - f_L$ (cf. Sec. 5.1) one notes the critical influence of the shape of the damping force $f_\eta(v)$ close above v^* on the time dependence of the vortex velocity $v(t)$. If the decrease of $f_\eta(v)$ above v^* is sufficiently slow, the time needed for any vortex to reach its terminal velocity will increase considerably. For negligible effective vortex masses on the order of a few quasi-particles per superconducting layer (obtained from free energy considerations cf. [Suhl65, Coff91, Soni98]) one might still expect negligible acceleration times. A recent theoretical work, however, has suggested the possibility of vortex masses equal to all particles inside the area of the vortex core for d-wave superconductors [Kopn98]. This dynamic mass arises from the inertia of excitations localized in the vortex core (which can far exceed the free-energy mass) and can be calculated from the force necessary to support an unsteady vortex motion

---

[42] All previous reports have always assumed a quasi-instantaneous rise (on experimental time scales) of the voltage [cf. Doet94].

[Kopn78]—a situation which matches the vortex dynamics at the instability. Such a vortex mass, increased by several orders of magnitude, should correspond to a substantially slower acceleration of the vortex system above the critical velocity and may result in voltage rise times on the observed time scales. However, an in-depth analysis based on this concept of the dynamic vortex mass would require a better knowledge about the precise behavior of the damping force above v^*.

Alternatively, vortex-vortex interactions occurring over an extended time scale could be the cause of the observed behavior. It has been suggested that a heavily entangled vortex liquid could exhibit such a high shear modulus that the dynamic response may occur on experimental time scales, in analogy with viscoelastic behavior in dense polymer melts [Nels89]. The long entanglement relaxation time could thus result in the observed extended vortex dynamic response at the critical current density. The particular appeal of the latter concept lies in phenomenological analogy to a glassy state, which was shown to be a prerequisite for the appearance of the instability in YBCO.

9 Concluding Remarks

The subject of the present work was the experimental investigation of vortex dynamics in the high-temperature superconductors $Bi_2Sr_2CaCu_2O_{8+\delta}$ and $YBa_2Cu_3O_7$. Special emphasis was placed on the influence of sample dimensionality and the resulting states of the flux-line system as well as on aspects of the interdependence of vortex dynamics at low and high vortex velocities and dissipation levels. At low vortex velocities the existence of pinning centers strongly influences the character of the dynamics. The relation between the energy barriers opposing flux-line activation on one side and the driving force and available thermal energy on the other side gives rise to a variety of activation mechanisms, and in the specific case of barriers diverging at low temperatures a vortex glass is observed. Yet, at high vortex velocities the influence of pinning becomes insubstantial and the flux-line dynamics are determined primarily by viscous drag opposing the driving force. As a result of the particular velocity dependence of the damping coefficient an instability in the dynamics of the flux-line system is observed.

The analysis of the activation energy in a series of BSCCO samples with systematically varied oxygen content revealed a clear dependence of the system's dimensionality in the low dissipative regime on the doping level arising from the influence of interstitial oxygen on interlayer coupling. Identified in the magnetic-field dependence of the activation energy and confirmed by the phase diagrams of the vortex glass analysis, a disentangled vortex state exists at low magnetic fields $B < B_e$ with the value of the entanglement field $B_e \sim 0.01$–0.1 T depending clearly on sample anisotropy. For more three-dimensional samples of higher oxygen concentration B_e increases systematically and the disentangled state, characterized by a 3D-like activation mechanism of single vortex or vortex bundle hopping, extends to considerably higher fields. This state, in which a three-dimensional vortex glass becomes possible below a glass temperature T_g, is diminished with decreasing oxygen content. In the opposite limit of high magnetic fields $B > B_{cr} \sim 1$ T all samples indicate the development of a two-dimensional pancake system resulting in a saturation of the activation energy and a magnetic-field independent glass temperature $T_g^{2D} \sim 0.3\, T_c$. The solid state below T_g^{2D} most likely consists of weakly coupled glass layers rather than a 3D vortex glass. At magnetic fields $B_e < B < B_{cr}$, between the three- and two-dimensional states, the dynamics of the flux-line system are dominated by two different competing flux-line activation mechanisms. For samples of higher oxygen content and anisotropy an increased contribution of the more 3D-like mechanism of flux-lattice shear was observed whereas in the reduced samples a double-kink mechanism of plastic lattice deformation prevails in this intermediate state.

In YBCO, on the other hand, at all magnetic fields the characteristic features of a 3D vortex glass transition were identified. CVC's taken at various magnetic fields and a wide range of temperatures closely followed the theoretical predictions of the VG model of Fisher, Fisher, and Huse and the extracted temperature dependence of the glass line $B_g \propto (1 - T_g/T_c)^{3/2}$ is in good agreement with prior reports. The observed magnetic-field dependence of the dynamic exponent z at low B also confirmed previous findings, but the onset of the increase in z was demonstrated to occur at considerably higher magnetic fields $\mu_0 H \sim 0.1$ T. Moreover, a clearly increased absolute value of the dynamic exponent $z \approx 9$ for $\mu_0 H \geq 0.1$ T was extracted from the analysis of the I–V glass isotherms, the crossover current density, the glass scaling, and the Vogel-Fulcher relation as opposed to theoretical estimates of $z = 4$–7 and published experimental results $z \approx 4$–6. The latter can be explained as a consequence of the limited electric-field range available in those measurements, as an artificial reduction of the data to a high electric-field range in the present experiment similarly allows to obtain z ranging from 6 to 9. However, the *complete* data covering more than seven orders of magnitude in E—available with the technique of extremely long measurement bridges developed in this work—are incompatible with $z \leq 6$ and unambiguously yield dynamic exponents $z \approx 9$. It is evident that the analysis of the vortex glass state is generally sensitive to the accessible electric-field window and a combination of different approaches including the close scrutiny of the crossover current is advisable. Thus, experimental influences such as finite size effects possibly obscuring the vortex glass transition and alternative models, including Bose glass and flux creep, could be excluded as potential origins of the high dynamic exponents. Instead, the qualitative behavior of all data is in full agreement with the vortex glass model. While a conclusive theoretical explanation is still lacking, the observed deviation of z may be a result of properties particular to the vortex system and not included in the theory of spin glasses from which the VG model is derived. Also, dynamic limitations of the flux-line ensemble may well influence the universality class and hence the value of the dynamic exponent while still preserving the underlying characteristics of this second order glass transition. Further investigations into this matter, in particular the refinement of the developed experimental technique and its application to a wider variety of samples and materials, will be useful.

In these low dissipative transport measurements the essential difference between the BSCCO and YBCO systems is their dimensionality and its effect on the formation of a vortex glass phase. The predominantly two-dimensional character of the flux-line system in BSCCO reappears at high vortex velocities, where a vortex instability was observed in the extended flux-flow regime. Arising from the non-monotonous velocity dependence of the viscous damping coefficient this instability was observed as a voltage jump in CVC's at high transport current densities. A quantitative analysis according to the theory of Larkin and Ovchinnikov and the extension by Bezuglyj and Shklovskij yielded the inelastic quasi-particle (electron-electron) scattering rates. The deviations of the instability parameters from the LO predictions were explained in terms of the influence of quasi-particle heating with a characteristic magnetic field B_T separating the two regimes of magnetic field where the pure LO mechanism of viscous drag and the effects of quasi-particle heating prevail. In the investigation of the anisotropy of this phenomenon with respect to the orientation of the magnetic field the data was well described by a quasi two-dimensional model in which only the magnetic-field component parallel to the c axis of the sample contributes to the mechanism of the instability.

Yet, in YBCO there exists no extended flux-flow region because of its three-dimensional character and the wide vortex glass regime. Similar voltage jumps were still observed in CVC's and the BS analysis allowed to extract the quasi-particle scattering rates, whose temperature dependence revealed a considerable influence of electron-phonon scattering at lower temperatures. More importantly, however, a direct dependence of the emergence of the instability on the vortex glass phase was discovered to exist throughout the entire phase diagram. Only for magnetic fields and temperatures within the glass phase results a discontinuous voltage jump in I-V curves whereas at higher B or T a gradual increase occurs. This distinct identity of the vortex glass line $B_g(T)$ and the jump line $B^*(T)$ was preserved for all YBCO films independent of sample quality and substrate material and could originate, for instance, from the depinning of the entire (glassy) vortex ensemble. A theoretical description of this correlation does not yet exist, but flux-line interactions characterizing the dynamics at low dissipative levels appear to be closely related to the high-dissipation mechanisms responsible for the vortex instability.

A further investigation of this phenomenon with rapid current pulses corroborated these results. Even though all discontinuous voltage jumps were shown to occur over an extended time interval, a definite distinction was still observed between the gradual voltage increase with current density in the fluid state and the sharp jump in the state below the vortex glass line. Whereas the gradual voltage increase corresponds to a continuous $E(J)$ dependence, in the glass phase the vortex instability manifests itself as a discontinuity in $E(J)$ at J^*. The observed rise time of the voltage jump on the order of a few microseconds is interpreted as the time span the vortex system needs to reach the dynamic state corresponding to the conditions above J^* after the current density has reached this critical value. Throughout the entire vortex glass phase the rise time, needed for the vortex ensemble to arrive at this state after the onset of the instability, is constant — independent of magnetic field and temperature. A possible explanation could be a limited flux-line acceleration due to large dynamic vortex masses. This notion could also explain the observed dependence of the rise time on the rate of increase in J: in shorter current pulses (of identical amplitude), where the current density and thus the driving force increase faster over time, the rise time is shortened. A different mechanism which may contribute to this process of a slow dynamic response of the vortex system is a long entanglement relaxation time. In light of these results, further theoretical and experimental work on the subjects of the vortex instability—particularly its investigation with an increased time resolution—and of the correlation between dynamics of low- and high-dissipative vortex phases are desirable.

Bibliography

[Abri57]   A.A. Abrikosov, *Zh. Éksperim. i Teor. Fiz.* **32**,1442 (1957) [*Sov. Phys. JETP* **5**, 1174 (1957)].

[Abul96]   Y. Abulafia *et al.*, *Phys. Rev. Lett.* **77**, 1596 (1996).

[Ande62]   P.W. Anderson, *Phys. Rev. Lett.* **9**, 309 (1962).

[Ande64]   P.W. Anderson and Y.B. Kim, *Rev. Mod. Phys.* **36**, 39 (1964).

[Ando92]   Y. Ando, H. Kubota, S. Tanaka, *Phys. Rev. Lett.* **69**, 2851 (1992).

[Ando93]   Y. Ando, H. Kubota, S. Tanaka, M. Aoyagi, H. Akoh, and S. Takada, *Phys. Rev.* **B47**, 5481 (1993).

[Anto99]   L. Antognazza, M. Decroux, N. Musolino, J.-M. Triscone, P. Reinert, E. Koller, S. Reymond, M. Chen, W. Paul, and Ø. Fischer to be published.

[Asla68]   L.G. Aslamasov and A.I. Larkin, *Phys. Lett.* **26A**, 238 (1968).

[Bale93]   G. Balestrino, M. Marinelli, E. Milani, A. Paoletti, and P. Paroli, *J. Appl. Phys.* **73**, 3903 (1995).

[Bale94]   G. Balestrino, D. V. Livanov, E. Milani, B. Camarota, D. Fiorani, and A. M. Testa, *Phys. Rev.* **B50**, 3446 (1994).

[Bale95]   G. Balestrino, A. Crisan, D.V. Livanov, E. Milani, M. Montuori, and A.A. Varlamov, *Phys. Rev.* **B51**, 9100 (1995).

[Bard56]   J. Bardeen, "Theory of Superconductivity," in S. Flügge (ed.), *Handbuch der Physik*, Vol. XV, Springer Verlag, Berlin (1956), p.303.

[Bard57]   J. Bardeen, L.N. Cooper, and J.R. Schrieffer, *Phys. Rev.* **108**, 1175 (1957).

[Bard65]   J. Bardeen and M.J. Stephen, *Phys. Rev.* **140**, A1197 (1965).

[Bass99a]   M. Basset, P. Voss-de Haan, G. Jakob, G. Wirth, and H. Adrian, to be published.

[Bass99b]   M. Basset, P. Voss-de Haan, F. Hilmer, G. Jakob, G. Wirth, and H. Adrian, to be published.

[Bedn86]   G. Bednorz and K.A. Müller, *Z. Phys.* **B64**, 189 (1986).

[Bezu92]   A.I. Bezuglyj and V.A. Shklovskij, *Physica C* **202**, 234 (1992).

[Bion56]   M.A. Biondi, M.P. Garfunkel, and A.O. McCoubrey, *Phys. Rev.* **102**, 1427 (1956).

125

[Blat94]   G. Blatter, M.V. Feigel'man, V.B. Geshkenbein, A.I. Larkin, and V.M. Vinokur, *Rev. Mod. Phys.* **66**, 1125 (1994).

[Boll91]   C.A. Bolle, P.L. Gammel, D.G. Grier, C.A. Murray, D.J. Bishop, D.B. Mitzi, and A. Kapitulnik, *Phys. Rev. Lett.* **66**, 112 (1991).

[Bonn93]   D.A Bonn *et al.*, *Phys. Rev.* **B47**, 11314 (1993).

[Bozo96]   I. Bozovic and J.N. Eckstein, "Superconductivity in Cuprate Superlattices" in *Physical Properties of High Temperature Superconductors*, Vol. V, ed. by D.M. Ginsberg (World Scientific, Singapore, 1996).

[Bran95]   E.H. Brandt, *Rep. Prog. Phys.* **58**, 1465 (1995).

[Brem59]   J.W. Bremer and V.L. Newhouse, *Phys. Rev.* **116**, 309 (1959).

[Caro64]   C. Caroli, P.G. de Gennes, and J. Matricon, *Phys. Lett.* **9**, 307 (1964).

[Char95]   M. Charalambous, R. H. Koch, T. Masselink, T. Doany, C. Feild, and F. Holtzberg, *Phys. Rev. Lett.* **75**, 2578 (1995).

[Chen98]   X.H. Chen, M. Yu, K.Q. Ruan, S.Y. Li, Z. Gui, G.C. Zhang, and L.Z. Cao, *Phys. Rev.* **B58**, 14219 (1998).

[Clem91]   J.R. Clem, *Phys. Rev.* **B43**, 7837 (1991).

[Coff91]   M. Coffey and Z. Hao, *Phys. Rev.* **B44**, 5230 (1991).

[Cohe97]   L.F. Cohen and H.J. Jensen, *Rep. Prog. Phys.* **60**, 1581 (1997).

[Copp90]   S.N. Coppersmith, M. Inui, and P. Littlewood, *Phys. Rev. Lett.* **64**, 2585 (1990).

[Cora54]   W.S. Corak, B.B. Goodman, C.B. Satterthwaite, and A. Wexler, *Phys. Rev.* **96**, 1442 (1954); **102**, 656 (1956).

[Crab97]   G.W. Crabtree and D.R. Nelson, *Physics Today*, 38 (April 1997).

[Dam99]   B. Dam, J.M. Huijbregtse, F.C. Klaassen, R.C.F. van der Geest, G. Doornbos, J.H. Rector, A.M. Testa, S. Freisem, J.C. Martínez, B. Stäuble-Pümpin, and R. Griessen, *Nature* **399**, 439 (1999).

[Daun46]   J.G. Daunt and K. Mendelssohn, *Proc. Roy. Soc.* (London) **A185**, 225 (1946).

[Deak93]   J. Deak, M.J. Darwin, and M. McElfresh, *Physica A* **200**, 332 (1993).

[Decr99]   M. Decroux, L. Antognazza, N. Musolino, J.M. Triscone, P. Reinert, E. Koller, S. Reymond, and Ø. Fischer, to be published in *Physica B* (1999).

[Dekk92a]   C. Dekker, W. Eidelloth, and R.H. Koch, *Phys. Rev. Lett.* **68**, 3347 (1992).

[Dekk92b]   C. Dekker, P.J.M. Wöltgens, R.H. Koch, B.W. Hussey, and A. Gupta, *Phys. Rev. Lett.* **69**, 2717 (1992).

[Deli97]   K. Deligiannis, P.A.J. de Groot, M. Oussena, S. Pinfold, R. Langan, R. Gagnon, and L. Taillefer, *Phys. Rev. Lett.* **79**, 2121 (1997).

[Doet94]   S.G. Doettinger, R. P. Huebener, R. Gerdemann, A. Kühle, A. Anders, T. G. Träuble, and J. C. Villégier, *Phys. Rev. Lett.* **73**, 1691 (1994).

[Doet95]   S.G. Doettinger, R.P. Huebener, and A. Kühle, *Physica C* **251**, 285 (1995).

[Emma92]   J.H.P. Emman, S.K.J. Lenzowski, J.H.J. Dalderop, and V.A.M. Brabers, *J. Cryst. Growth* **118**, 477 (1992).

[Farr89]   D.E. Farrell, S. Bonham, J. Foster, Y.C. Chang, P.Z. Jiang, D.J. Lam, K.G. Vandervoort, and V.G. Kogan, *Phys. Rev. Lett.* **63**, 782 (1989).

[Farr90]   D.E. Farrel, R.G. Beck, M.F. Booth, C.J. Allen, E.D. Bukowski, and D.M. Ginsberg, *Phys. Rev.* **B42**, 6758 (1990).

[Feig89]   M.V. Feigel'man, V.B. Geshkenbein,, A.I. Larkin, and V.M. Vinokur, *Phys. Rev. Lett.* **63**, 2303 (1989).

[Fish89a]   M.P.A. Fisher, *Phys. Rev. Lett.* **62**, 1415 (1989).

[Fish89b]   M.P.A. Fisher, P.B. Weichman, G. Grinstein, and D.S. Fisher, *Phys. Rev.* **B40**, 546 (1989).

[Fish91]   D.S. Fisher, M.P.A. Fisher, and D.A. Huse, *Phys. Rev.* **B43**, 130 (1991).

[Gamm88]   P.L. Gammel, L.F. Schneemeyer, J.V. Waszczak, and D.J. Bishop, *Phys. Rev. Lett.* **61**, 1666 (1988).

[Gamm90]   P.L. Gammel, *J. Appl. Phys.* **67**, 4676 (1990).

[Gamm91]   P.L. Gammel, L.F. Schneemeyer, and D.J. Bishop, *Phys. Rev. Lett.* **66**, 953 (1991).

[Gao93]   F. Gao *et al.*, *Appl. Phys. Lett.* **63**, 2274 (1993).

[Genn64]   P.G. de Gennes and J. Matricon, *Rev. Mod. Phys.* **36**, 45 (1964).

[Genn66]   P.G. de Gennes, *Superconductivity of Metals and Alloys*, W.A. Benjamin, New York, 1966, reprinted by Addison-Wesley, Reading, MA, 1989.

[Gesh89]   V. Geshkenbein, A. Larkin, M. Feigel'man, V. Vinokur, *Physica C* **162–164**, 239 (1989).

[Giam96]   T. Giamarchi and P. Le Doussal, *Phys. Rev. Lett.* **76**, 3408 (1996).

[Ginz50]   V.L. Ginzburg and L.D. Landau, *Zh. Éksperim. i Teor. Fiz.* **20**, 1064 (1950).

[Ginz53]   V.L. Ginzburg, *Fortschr. Phys.* **1**, 101 (1953).

[Glaz91]   L.I. Glazman and A.E. Koshelev, *Phys. Rev.* **B43**, 2835 (1991).

[Glov56]   R.E. Glover and M. Tinkham, *Phys. Rev.* **104**, 844 (1956); **108**, 243 (1957).

[Goll94]   F. Gollnik *et al.*, *Physica C* **235–240**, 1933 (1994).

[Gork73]   L.P. Gor'kov and N.B. Kopnin, *Sov. Phys. JETP* **37**, 183 (1973).

[Goup97]   C. Goupil, A. Ruyter, V. Hardy, and Ch. Simon, *Physica C* **278**, 23 (1997).

[Grie90]   R. Griessen, *Phys. Rev. Lett.* **64**, 1674 (1990).

[Gupt93]   S.K. Gupta, P. Berdahl, R.E. Russo, G. Briceño, and A. Zettl, *Physica C* **206**, 335 (1993) and references therein.

[Gure87]   A.Vl. Gurevich and R.G. Mints, *Rev. Mod. Phys.* **59**, 941 (1987).

[Haib96]   P. Haibach, Diplomarbeit, TH Darmstadt (1996).

[Halp79]   B.I. Halperin and D.R. Nelson, *J. Low Temp. Phys.* **36**, 599 (1979).

[Hara93]   K. Harada, T. Matsuda, H. Kasai, J. E. Bonevich, T. Yoshida, U. Kawabe, and A. Tonomura, *Phys. Rev. Lett.* **71**, 3371 (1993).

[Harl95]   D.J. Van Harlingen, *Rev. Mod. Phys.* **67**, 515 (1995).

[Hemp64]   C.F. Hempstead and Y.B. Kim, *Phys. Rev. Lett.* **12**, 145 (1964).

[Hend96]   W. Henderson, E.Y. Andrei, M.J. Higgins, and S. Bhattacharya, *Phys. Rev. Lett.* **77**, 2077 (1996).

[Hett89]   J.D. Hettinger, A.G. Swanson, W.J. Skocpol, J.S. Brooks, J.M. Graybeal, P.M. Mankiewich, R.E. Howard, B.L. Straughn, and E.G. Burkhardt, *Phys. Rev. Lett.* **62**, 2044 (1989).

[Horo98]   B. Horovitz and T.R. Goldin, *Phys. Rev. Lett.* **80**, 1734 (1998).

[Hou97]   L. Hou, J. Deak, P. Metcalf, M. McElfresh, and G. Preosti, *Phys. Rev.* **B55**, 11806 (1997).

[Iye92]   Y. Iye *et al.*, *Physica* (Amsterdam) **199C**, 154 (1992).

[Jako99]   G. Jakob, P. Voss-de Haan, M. Wagner, Z.L. Xiao, and H. Adrian, accepted for publication in *Physica B* (1999).

[Jose62]   B.D. Josephson, *Phys. Lett.* **1**, 251 (1962); *Adv. Phys.* **14**, 419 (1965).

[Kame11]   H. Kamerlingh Onnes, *Leiden Comm.* **120b**, **122b**, **124c** (1911).

[Kes89]   P.H. Kes, J. Aarts, J. van den Berg, C.J. van der Beek, and J.A. Mydosh, *Supercond. Sci. Technol.* **1**, 242 (1989).

[Kes90]   P.H. Kes, J. Aarts, V.M. Vinokur, and C.J. van der Beek, *Phys. Rev. Lett.* **64**, 1063 (1990).

[Kes96]   P.H. Kes, *J. Phys. I* (France) **6**, 2327 (1996).

[Khay96]   B. Khaykovich, E. Zeldov, D. Majer, T.W. Li, P.H. Kes, and M. Konczykowski, *Phys. Rev. Lett.* **76**, 2555 (1996).

[Khay97]   B. Khaykovich, M. Konczykowski, E. Zeldov, R.A. Doyle, D. Majer, P.H. Kes, and T.W. Li, *Phys. Rev.* **B56**, R517 (1997).

[Kim63]   Y.B. Kim, C.F. Hempstead, and A.R. Strnad, *Phys. Rev.* **131**, 2486 (1963).

[Kirt95]   J.R. Kirtley, C.C. Tsuei, J.Z. Sun, C.C. Chi, L.S. YuJahnes, A. Gupta, M. Rupp, and M.B. Ketchen, *Nature* **373**, 225 (1995).

[Klei85]   W. Klein, R.P. Huebener, S. Gauss, and J. Parisi, *J. Low Temp. Phys.* **61**, 413 (1985).

[Klug95]   T. Kluge, Y.Koike, A. Fujiwara, M. Kato, T. Noji, and Y. Saito, *Phys. Rev.* **B52**, R727 (1995).

[Koch89]   R.H. Koch, V. Foglietti, W.J. Gallagher, G. Koren, A. Gupta, and M.P.A. Fisher, *Phys. Rev. Lett.* **63**, 1511 (1989).

[Koch90]   R.H. Koch, V. Foglietti, and M.P.A. Fisher, *Phys. Rev. Lett.* **64**, 2586 (1990).

[Koga89]   V. Kogan and L.J. Campbell, *Phys. Rev. Lett.* **62**, 1552 (1989).

[Kopn78]   N.B. Kopnin, *Pis'ma Zh. Éksperim. Teor. Fiz.* **27**, 417 (1978) [*JETP Lett.* **27**, 390 (1978)].

[Kopn98]   N.B. Kopnin and V.M. Vinokur, *Phys. Rev. Lett.* **81**, 3952 (1998).

[Kosh94]   A.E. Koshelev and V.M. Vinokur, *Phys. Rev. Lett.* **73**, 3580 (1994).

[Koss98]   W.J. Kossler, Y. Dai, K.G. Petzinger, A.J. Greer, D.Ll. Williams, E. Koster, D.R. Harshman, and D.B. Mitzi, *Phys. Rev. Lett.* **80**, 592 (1998).

[Kost73]   J.M. Kosterlitz and D.J. Thouless, *J. Phys.* **C6**, 1181 (1973).

[Kota94]   Y. Kotaka *et al.*, *Physica* (Amsterdam) **235C**, 1529 (1994).

[Kötz94a]   J. Kötzler, M. Kaufmann, G. Nakielski, R. Behr, and W. Assmus, *Phys. Rev. Lett.* **72**, 2081 (1994).

[Kötz94b]   J. Kötzler, G. Nakielski, M. Baumann, R. Behr, F. Goerke, and E.H. Brandt, *Phys. Rev.* **B50**, 3384 (1994).

[Kräm98]   A. Krämer, *Physica C* **309**, 33 (1998).

[Kuce92]   J.T. Kucera, T.P. Orlando, G. Virshup, and J.N. Eckstein, *Phys. Rev.* **B46**, 11004 (1992).

[Lark70]   A.I. Larkin, *Zh. Éksperim. Teor. Fiz.* **58**, 1466 (1970).

[Lark71]   A.I. Larkin and Yu.N. Ovchinnikov, *Sov. Phys. JETP* **37**, 557 (1971).

[Lark75]   A.I. Larkin and Yu.N. Ovchinnikov, *Zh. Éksperim. Teor. Fiz.* **68**, 1915 (1975) [*Sov. Phys. JETP* **41**, 960 (1976)].

[Lark79]   A.I. Larkin and Yu.V. Ovchinnikov, *J. Low Temp. Phys.* **34**, 409 (1979).

[Lawr70]   W.E. Lawrence and S. Doniach, in *Proc. of the 12th Int. Conf. on Low Temp. Phys.*, Kyoto, Japan, 1970, ed. by E. Kanda, [Keigaku, Tokyo, 1971], p.361.

[Legr96]   L. Legrand, I. Rosenman, R. G. Mints, G. Collin, and E. Janod, *Europhys. Lett.* **34**, 287 (1996).

[Lind10]   F. Lindemann, *Phys. Z.* (Leipzig) **11**, 69 (1910).

[Lond35]   F. and H. London, *Proc. Roy. Soc.* (London) **A149**, 71 (1935).

[Lope96a]   D. López, E.F. Righi, G. Nieva, F. de la Cruz, W.K. Kwok, J. A. Fendrich, G.W. Crabtree, and L. Paulius, *Phys. Rev.* **B53**, 8895 (1996).

[Lope96b]   D. López, E.F. Righi, G. Nieva, F. de la Cruz, *Phys. Rev. Lett.* **76**, 4034 (1996).

[Lund98]   B. Lundqvist, J. Larsson, A. Herting, Ö. Rapp, M. Andersson, Z. G. Ivanov, and L.-G. Johansson, *Phys. Rev.* **B58**, 6580 (1998).

[Ma91]   H.-r. Ma and S.T. Chui, *Phys. Rev. Lett.* **67**, 505 (1991).

[Maed88]   H. Maeda, Y. Tanaka, M. Fukutomi, and T. Asano, *Jpn. J. Appl. Phys.* **27**, L209 (1988).

[Maki71]   K. Maki and H. Takayama, *Prog. Theor. Phys.* **26**, 1651 (1971).

[Male90]   M.P. Maley, J.O. Willis, H. Lessure, and M.E. McHenry, *Phys. Rev.* **B42**, 2639 (1990).

[Mars93]   C.D. Marshall, A. Tokmakoff, I.M. Fishman, C.B. Eom, J.M. Phillips, and M.D. Fayer, *J. Appl. Phys.* **73**, 850 (1993).

[Mart92]   J.C. Martínez, S.H. Brongersma, A. Koshelev, B. Ivlev, P.H. Kes, R.P. Griessen, D.G. de Groot, Z. Tarnavski, and A.A. Menovsky, *Phys. Rev. Lett.* **69**, 2276 (1992).

[Meis33]   W. Meissner and R. Ochsenfeld, *Naturwissenschaften* **21**, 787 (1933).

[Mich92]   P.C. Michael, J.U. Trefny, and B. Yarar, *J. Appl. Phys.* **72**, 107 (1992).

[Mitz90]   D.B. Mitzi, C.W. Lombardo, A. Kapitulnik, S.S. Laderman, and R.D. Jacowitz, *Phys. Rev.* **B41**, 6564 (1990).

[Miu98]   L. Miu, G. Jakob, P. Haibach, Th. Kluge, U. Frey, P. Voss-de Haan, and H. Adrian, *Phys. Rev.* **B57**, 3144 (1998).

[Musi80]   L.E. Musienko, I.M. Dmitrenko, and V.G. Volotskaya, *JETP Lett.* **31**, 567 (1980).

[Nahu91]   M. Nahum, S. Verghese, P.L. Richards, and K. Char, *Appl. Phys. Lett.* **59**, 2034 (1991).

[Nels77]   D.R. Nelson and J.M. Kosterlitz, *Phys. Rev. Lett.* **39**, 1201 (1977).

[Nels89]   D.R. Nelson and H.S. Seung, *Phys. Rev.* **B39**, 9153 (1989).

[Nels92]   D.R. Nelson and V.M. Vinokur, *Phys. Rev. Lett.* **68**, 2398 (1992).

[Nels93]   D.R. Nelson and V.M. Vinokur, *Phys. Rev.* **B48**, 13060 (1993).

[Newn93]   D.M. Newns, H.R. Krishnamurthy, P.C. Pattnaik, C.C. Tsuei, C.C. Chi, and C.L. Cane, *Physica* (Amsterdam) **186–188B**, 801 (1993).

[Noji96]   T. Nojima, A. Kakinuma, Y. Kuwasawa, *Physica C* **263**, 424 (1996).

[Nuss91]   M.C. Nuss, P.M. Mankiewich, M.L. O'Malley, E.H. Westerwick, and P.B. Littlewood, *Phys. Rev. Lett.* **66**, 3305 (1991).

[Pals89]   T.T.M. Palstra, B. Batlogg, R.B. van Dover, L.F. Schneemeyer, and J.V. Waszczak, *Appl. Phys. Lett.* **54**, 763 (1989).

[Pals90]   T.T.M. Palstra, B. Batlogg, R.B. van Dover, L.F. Schneemeyer, and J.V. Waszczak, *Phys. Rev.* **B41**, 6621 (1990).

[Paul99]   M. Pauly, R. Ballou, G. Fillion, and J.-C. Villégier, to be published in *Physica B* (1999).

[Pick92]   W.E. Pickett, H. Krakauer, R.E. Cohen, and D.J. Singh, *Science* **255**, 46 (1992).

[Pla91]   O. Pla and F. Nori, *Phys. Rev. Lett.* **67**, 919 (1991).

[Pres93]   M.R. Presland, J.L. Tallon, R.G. Buckley, R.S. Liu, and N.E. Flower, *Physica C* **176**, 95 (1993).

[Quin94]   S.M. Quinlan, D.J. Scalapino, and N. Bulut, *Phys. Rev.* **B49**, 1470 (1994).

[Reed73]   R.E. Reed-Hill, *Physical Metallurgy Principles*, 2nd ed. (V. Nostrand, New York, 1973), p.848.

[Robe94]   J.M. Roberts, B. Brown, B.A. Hermann, and J. Tate, *Phys. Rev.* **B49**, 6890 (1994).

[Ruck97]   B.J. Ruck, J.C. Abele, H.J. Trodahl, S.A. Brown, and P. Lynam, *Phys. Rev. Lett.* **78**, 3378 (1997).

[Safa92]   H. Safar, P.L. Gammel, D.J. Bishop, *Phys. Rev. Lett.* **68**, 2672 (1992).

[Samo95]   A.V. Samoilov, M. Konczykowski, N.-C. Yeh, S. Berry, and C.C. Tsuei, *Phys. Rev. Lett.* **75**, 4118 (1995).

[Sawa98]   A. Sawa, H. Yamasaki, Y. Mawatari, H. Obara, M. Umeda, and S. Kosaka, *Phys. Rev.* **B58**, 2868 (1998).

[Schi93]   A. Schilling, M. Cantoni, J.D. Guo, and H.R. Ott, *Nature* **363**, 56 (1993).

[Seng91]   S. Sengupta, C. Dasgupta, H.R. Krishnamurthy, G.I. Menon, and T.V. Ramakrishnan, *Phys. Rev. Lett.* **67**, 3444 (1991).

[Shen88]   Z.Z. Sheng and A.M. Hermann, *Nature* **332**, 55 (1988).

[Silv93]   R.M. Silver, A.L. de Lozanne, and M. Thompson, *IEEE Trans. Appl. Superconductivity* **3**, 1394 (1993).

[Silv97]   E. Silva, S. Sarti, M. Guira, R. Fastampa, and R. Marcon, *Phys. Rev.* **B55**, 11115 (1997).

[Skoc74]   W.J. Skocpol, M.R. Beasley, and M. Tinkham, *J. Appl. Phys.* **45**, 4054 (1974).

[Skoc75]   W.J. Skocpol, M. Tinkham, *Rep. Prog. Phys.* **38**, 1049 (1975).

[Soni98]   E.B. Sonin, V.B. Geshkenbein, A. van Otterlo, and G. Blatter, *Phys. Rev.* **B57**, 575 (1998).

[Stei94]   F. Steinmeyer, R. Kleiner, P. Müller, H. Müller, and K. Winzer, *Europhys. Lett.* **25**, 459 (1994).

[Sudb91]   A. Sudbø and E.H. Brandt, *Phys. Rev.* **B43**, 10482 (1991); *Phys. Rev. Lett.* **66**, 1781 (1991).

[Sude98]   Y.S. Sudershan, A. Rastogi, S.V. Bhat, A.K. Grover, Y. Yamaguchi, K. Oka, and Y. Nishihara, *Supercond. Sci. Technol.* **11**, 1372 (1998).

[Suhl65]   H. Suhl, *Phys. Rev. Lett.* **14**, 226 (1965).

[Sun89]   J.Z. Sun, K. Char, M.R. Hahn, T.H. Geballe, and A. Kapitulnik, *Appl. Phys. Lett.* **54**, 663 (1989).

[Tall95]   J.L. Tallon, C. Bernhard, H. Shaked, R.L. Hittermann, and J.D. Jorgenson, *Phys. Rev.* **B51**, 12911 (1995).

[Tink64]   M. Tinkham, *Phys. Rev. Lett.* **13**, 804 (1964).

[Tink88]   M. Tinkham, *Phys. Rev. Lett.* **61**, 1658 (1988).

[Tink96]   M. Tinkham, *Introduction to superconductivity*, 2nd ed., McGraw-Hill, New York, 1996.

[Toul70]   Y.S. Touloukian, R.W. Powell, C.Y. Ho, and P.G. Klemens, *Thermophysical Properties of Matter* (IFI/Plenum, New York, 1970), Vol. 2, p. 166.

[Tsue97]   C.C. Tsuei, J.R. Kirtley, Z.F. Ren, J.H. Wang, H. Raffy, and Z.Z. Li, *Nature* **387**, 481 (1997).

[Vino90a]   V.M. Vinokur, P.H. Kes, and A.E. Koshelev, *Physica C* **168**, 29 (1990).

[Vino90b]   V.M. Vinokur, M.V. Feigel'man, V.B. Geshkenbein, and A.I. Larkin, *Phys. Rev. Lett.* **65**, 259 (1990).

[Voss99]   P. Voss-de Haan, P. Haibach, G. Jakob, and H. Adrian, to be published in *Physica C*.

[WagM98]   M. Wagner, Diplomarbeit, Mainz (1998).

[Wagn91]   P. Wagner, Diplomarbeit, TH Darmstadt (1991).

[Wagn93]   P. Wagner, F. Hillmer, U. Frey, H. Adrian, T. Steinborn, L. Ranno, A. Elschner, I. Heyvaert, and Y. Bruynseraede, *Physica C* **215**, 123 (1993).

[Wagn94a]   P. Wagner, Dissertation, TH Darmstadt (1994).

[Wagn94b]   P. Wagner, F. Hillmer, U. Frey, and H. Adrian, *Phys. Rev.* **B49**, 13184 (1994).

[Wagn95]   P. Wagner, U. Frey, F. Hillmer, and H. Adrian, *Phys. Rev.* **B51**, 1206 (1995).

Bibliography

[Wang92]   Z.D. Wang and C.S. Ting, *Phys. Rev.* **B46**, 284 (1992).

[Wang98]   Z.H. Wang, *Physica C* **306**, 253 (1998).

[Wen97]   H.-h. Wen, X.X. Yao, R.L. Wang, H.C. Li, S.Q. Guo, and Z.X. Zhao, *Physica C* **282–287**, 351 (1997).

[Wen98]   H.-h. Wen, P. Ziemann, H.A. Radovan, and S.L. Yan, *Europhys. Lett.* **42**, 319 (1998).

[Whit91]   W.R. White, A. Kapitulnik, and M.R. Beasley, *Phys. Rev. Lett.* **66**, 2826 (1991).

[Wink96]   L. Winkeler, S. Sadewasser, B. Beschoten, H. Frank, F. Nouvertne, and G. Guntherodt, *Physica C* **265**, 194 (1996).

[Wölt93]   P.J.M. Wöltgens, C. Dekker, J. Swüste, and H.W. de Wijn, *Phys. Rev.* **B48**, 16826 (1993).

[Wölt95]   P.J.M. Wöltgens, C. Dekker, R.H. Koch, B.W. Hussey, and A. Gupta, *Phys. Rev.* **B52**, 4536 (1995).

[Wort88]   T.K. Worthington *et al.*, *Physica* **153C**, 32 (1988).

[Wu87]   M.K. Wu, J.R. Ashburn, C.J. Torng, P.H. Hor, R.L. Meng, L. Gao, Z.J. Huang, Y.Q. Wang, and C.W. Chu, *Phys. Rev. Lett.* **58**, 908 (1987).

[Xeni93]   D.G. Xenikos, J.T. Kim, and T.R. Lemberger, *Phys. Rev.* **B48**, 7742 (1993).

[Xiao96]   Z.L. Xiao and P. Ziemann, *Phys. Rev.* **B53**, 15265 (1996).

[Xiao97]   Z.L. Xiao and P. Ziemann, *Z. Phys. B* **104**, 451 (1997).

[Xiao98a]   Z.L. Xiao, P. Voss-de Haan, G. Jakob, and H. Adrian, *Phys. Rev.* **B57**, R736 (1998).

[Xiao98b]   Z.L. Xiao, E.Y. Andrei, and P. Ziemann, *Phys. Rev.* **B58**, 11185 (1998).

[Xiao99]   Z.L. Xiao, P. Voss-de Haan, G. Jakob, Th. Kluge, P. Haibach, H. Adrian, and E.Y. Andrei, *Phys. Rev.* **B59**, 1481 (1999).

[Yama94]   Yamasaki, *Phys. Rev.* **B50**, 12959 (1994).

[Yasu93]   T. Yasuda, S. Takano, and L. Rinderer, *Physica C* **208**, 385 (1993).

[Yesh88]   Y. Yeshurun and A.P. Malozemoff, *Phys. Rev. Lett.* **60**, 2202 (1988).

[Yu92]   R.C. Yu, M.B. Salamon, J.P. Lu, and W.C. Lee, *Phys. Rev. Lett.* **69**, 1431 (1992).

[Yu97]   Y. Yu, Z.Y. Zheng, M.J. Qin, X.X. Yao, L.P. Ma, H.C. Li, and L.Li, *Physica C* **292**, 197 (1997).

[Zhan93]   H. Zhang and H. Sato, *Phys. Rev. Lett.* **70**, 1697 (1993).

Publications

The following publications were produced in relation to this work:

A. Wienss, G. Jakob, P. Voss-de Haan, and H. Adrian "History dependence of the magnetization of thin HTSC films — An explanation for distorted SQUID signals." *Physica C* **280**, 158 (1997).

L. Miu, G. Jakob, P. Haibach, Th. Kluge, U. Frey, P. Voss-de Haan, and H. Adrian "Length-scale-dependent vortex-antivortex unbinding in epitaxial $Bi_2Sr_2CaCu_2O_{8+\delta}$ films." *Phys. Rev.* **B57**, 3144 (1998).

Z.L. Xiao, P. Voss-de Haan, G. Jakob, and H. Adrian "Voltage jumps in current-voltage characteristics in $Bi_2Sr_2CaCu_2O_{8+\delta}$ superconducting films: Evidence for flux-flow instability under the influence of self-heating." *Phys. Rev.* **B57**, R736 (1998).

Z.L. Xiao, P. Voss-de Haan, G. Jakob, Th. Kluge, P. Haibach, H. Adrian, and E.Y. Andrei "Flux-flow instability and its anisotropy in $Bi_2Sr_2CaCu_2O_{8+\delta}$ superconducting films." *Phys. Rev.* **B59**, 1481 (1999).

P. Voss-de Haan, G. Jakob, and H. Adrian "High dynamic exponents in vortex glass transitions: Dependence of critical scaling on the electric-field range." *Phys. Rev.* **B60**, 12443 (1999).

G. Jakob, P. Voss-de Haan, M. Wagner, Z. Xiao, and H. Adrian "Flux-flow instability and heating effects in $Bi_2Sr_2CaCu_2O_8$ and $YBa_2Cu_3O_7$ thin films." Accepted for publication in *Physica B*.

P. Voss-de Haan, G. Jakob, and H. Adrian "Correlation of the voltage instability and the vortex glass phase in $YBa_2Cu_3O_7$." To be published in *Phys. Rev. Lett.*

P. Voss-de Haan, M. Wagner, G. Jakob, and H. Adrian "Time evolution of the vortex instability in $YBa_2Cu_3O_7$ films." To be published in *Europhys. Lett.*

P. Voss-de Haan, P. Haibach, G. Jakob, and H. Adrian "Variation of vortex tension and dimensionality with oxygen content in epitaxial $Bi_2Sr_2CaCu_2O_{8+\delta}$ films." To be published in *Physica C*.

M. Basset, P. Voss-de Haan, G. Jakob, G. Wirth, and H. Adrian "Bose glass analysis with high electric-field sensitivity on extremely long measurement bridges." To be published in *Phys. Rev.* **B**.

M. Basset, P. Voss-de Haan, F. Hilmer, G. Jakob, G. Wirth, and H. Adrian "Influence of power dissipation and Joule heating in transport measurements at high current densities." To be published in *Physica C.*

Acknowledgements

The last few years and the work that has filled up a good part of it (the results of which you may have just read) have been at the same time challenging and interesting, stressful and enjoyable. Many people have contributed—except maybe to the stressful part, which was mostly machine- and self-made. It has been more than a valuable experience.

First of all, I would like to thank Prof. Dr. H. Adrian for giving me the opportunity to complete this PhD thesis in his work group, for always being interested in the experiments and analyses, and for providing all the support a PhD student can ask for. I thoroughly enjoyed this experience in condensed matter physics, a field of physics that was quite new to me when I began this work.

I am deeply grateful to Dr. Gerhard Jakob, the head of our vortex-dynamics group, for many useful and enlightening discussions. My work has probably benefited more than I can estimate from his knowledge and understanding of the 'vortex matter'.

Also, I wish to thank Dr. Michael Huth and Dr. Juan Carlos Martínez, who also have always been very helpful and supportive in every way.

Dr. Zhili L. Xiao started the work on vortex instabilities in our group and without his initiative I probably would never have pursued those topics that make up the second part of this thesis.

I am indebted to my former Diploma-student Dipl.-Phys. Michael Wagner for the preparation of several samples and his experimental support in the investigation of the instability in YBCO and to Dipl.-Phys. Michael Basset, in particular for lending a helping hand during the last stages of experimental work and for continuing some of the projects.

For their good company, the interesting discussions, the never-ending entertainment, and their good sense of humor I am thankful to my colleagues, Dipl.-Phys. Patrick Haibach (who also prepared the invaluable oxygen-annealed BSCCO samples), Dr. Christoph Schwan, Dipl.-Ing. Holger Meffert, Dr. Frank Hillmer, Dr. Martin Jourdan, Dr. Thom Kluge, and all other members of the AG Adrian.

Furthermore, I am grateful to Prof. V. M. Vinokur and Prof. V. A. Shklovskij for a number of valuable discussions and to Prof. Douglas D. Osheroff for his inspiration and support some time ago.

Neither last nor least, I have to thank Mr. E. Griess and Mr. H. Geibel for their never-ending supply of liquid Helium and Mr. P. Becker and his co-workers from the machine shop for their help.

Yet most of all, I wish to thank my mother Antoinette Voss-de Haan and Sandy Mohr, who have helped me arrive at this point, and my father, who I wish would have gotten a chance to read this.

This work was supported by the "Sonderforschungsbereich 262, Glaszustand und Glasübergang nichtmetallischer amorpher Materialien" of the Deutsche Forschungsgemeinschaft (DFG).

135

Index

www.ingramcontent.com/pod-product-compliance
Lightning Source LLC
Chambersburg PA
CBHW061325220326
41599CB00026B/5033